Cathérine Meißner

High-resolution sensitivity studies with the regional climate model COSMO-CLM

Wissenschaftliche Berichte des Instituts für Meteorologie und Klimaforschung der Universität Karlsruhe (TH)
Band 42

Herausgeber: Prof. Dr. Ch. Kottmeier

Institut für Meteorologie und Klimaforschung der Universität Karlsruhe (TH) (gemeinsam betrieben mit dem Forschungszentrum Karlsruhe)
Kaiserstr. 12, 76128 Karlsruhe

High-resolution sensitivity studies with the regional climate model COSMO-CLM

by
Cathérine Meißner

universitätsverlag karlsruhe

Dissertation, Universität Karlsruhe (TH),
Fakultät für Physik, 2008
Referenten: Prof. Dr. Christoph Kottmeier, Prof. Dr. Sarah Jones

Impressum

Universitätsverlag Karlsruhe
c/o Universitätsbibliothek
Straße am Forum 2
D-76131 Karlsruhe
www.uvka.de

Universitätsverlag Karlsruhe 2008
Print on Demand

ISSN: 0179-5619
ISBN: 978-3-86644-228-3

Abstract

High-resolution regional ensemble climate simulations with the regional climate model COSMO-CLM are performed for Southwest Germany to study the sensitivity of meteorological and hydrological variables to simulation set-up, including the domain size, driving data, horizontal resolution and physical parameterisations and parameter settings. The model setup found adequate for such simulations is a domain including the Alps, ERA-40 reanalysis data as driving data, and a horizontal resolution of 7 km. The sensitivity of simulation results to a changed model setup is highest for the change in driving data and is higher in winter than in summer.

This adequate model setup is used to investigate the influence of the land surface scheme on COSMO-CLM simulations. The standard land surface scheme TERRA_LM is replaced by the land surface scheme VEG3D, which contains an explicit vegetation layer. Stand-alone simulations with both land surface schemes show better agreement with observations for the VEG3D scheme, especially over high vegetation. Coupled online with the COSMO-CLM, both schemes yield similar results on the spatial patterns of the meteorological variables but the absolute values may differ considerably. No model system gives better results than the other for 2m-temperature and precipitation compared to observations, and the difference in TERRA_LM and VEG3D simulation is similar to the difference obtained by changing other physical parameterisations or the time-integration scheme. Freezing and melting processes in the soil are implemented in VEG3D to make the scheme applicable for climate simulations. Stand-alone simulations with the new scheme yield better results than those without the consideration of freezing and melting processes. Better results in stand-alone simulations are obtained when using different soil types within one soil column, instead of using one single soil type for the whole column. Therefore, a soil type inventory for Germany for the coupled model system COSMO-CLM/VEG3D is provided in this thesis, which contains several soil types within one soil column.

A strategy for a statistical-dynamical downscaling method is developed and evaluated to replace time consuming day by day simulations. The method shows the ability to yield results similar to those of the continuous simulation.

CONTENTS

1 INTRODUCTION

The Fourth Assessment Report of Intergovernmental Panel on Climate Change (IPCC) pointed out that the observed warming of the climate system is unequivocal and that there is no chance to stop the climate change within the coming decades (ALLEY et al., 2007). Even if the concentration of all greenhouse gases were frozen to the level of the year 2000, a further warming of about 0.1°C per decade is to be expected. Therefore, it is necessary to estimate the consequences climate change will have on regional climate and to develop strategies on how to adapt to the climate change caused by increasing global warming. As a basis for developing plans for adaptation and mitigation on a regional scale, it is necessary to predict the changes of climate variables, like temperature and precipitation, and their statistics over the coming decades, and to determine the uncertainties in such predictions. The demand for reliable climate simulations for specific regions is increasing.

Global models are not able to resolve complex topographies due to their coarse resolution of more than 100 km. Hence, the forecasts of global models are not suitable for climate impact studies in regions with complex topography like Southwest Germany (KLIWA, 2006). Regional climate models with higher resolution than global models are used to downscale the results of global climate models onto a finer grid to provide reliable results for such regions. This can be done for limited areas only, due to the excessive computing time, which has been a limiting factor in climate prediction up to now.

Due to their higher resolution, regional climate models are expected to give better results than global models (MO et al., 2000), especially for extreme events (CHRISTENSEN et al., 1998, WANG et al., 2003). Extreme values of troughs, intense precipitation, and strong winds tend to be better simulated by regional models than by global models (GIORGI and MEARNS, 1999). Their precipitation differences to global climate models mainly arise from orographic forcing and rain shadowing effects (GIORGI et al., 1994, JONES et al., 1997, LEUNG et al., 2004).

Due to the high demand for reliable regional climate simulations, some new regional climate models have been developed in the recent past, mainly from existing weather prediction models. For example the regional climate model COSMO-CLM (BÖHM et al., 2006), which is used for all the studies presented in this thesis. This model was developed from the existing weather prediction model Lokal-Modell (LM) (DOMS et al., 2005) from the German Weather Service (DWD). This work was mainly done by researchers from the Potsdam Institute of Climate Impact Research (PIK), the University of Cottbus, the GKSS Research Centre, and DWD. Forschungszentrum Karlsruhe (Institute of Meteorology and Climate Research, IMK-TRO) joined the consortium in 2004 and focusses its efforts on model developments towards highest resolution, non-hydrostatic simulations, and land-surface schemes. Today, the CLM community comprises researchers from about 20 different institutes using the model for climate studies and contributing to its further development. The model has become the community model for German climate research on the regional scale. In 2007, the model was combined with LM into one source code. From model version 4.0 onwards, the model is renamed from CLM to COSMO-CLM. The advantage over other climate models is that COSMO-CLM is a "living" climate model. This means that the source code is permanently improved and new parameterisations are developed, as the source code is identical to the source code of the COSMO model of DWD used for the operational numerical weather forecast. Therefore, experience with the weather forecast version can be transferred to the climate version and vice versa. One difference between weather forecast and climate prediction with the same model is the assimilation of observed data to correct the weather forecast. Prognostic climate model runs needs to be performed without any correction by observations. Another difference is the update of vegetation parameters during the annual cycle, which is not necessary for weather forecast models that only run over several days. The initialisation of the model is more important in weather forecasting than in climate mode. Climate models are given some time to adapt and to weight out imbalances that occur due to the initialisations. This is not possible in weather forecasting and the initialisation of the model is important for the quality of the results.

Over the last decade, the horizontal resolution of regional climate models was about 50 km. However, a horizontal resolution of 50 km is still too coarse for climate change investigations that concern hydrology or water management in orographically structured regions (CHRISTENSEN et al., 1998). Effects caused by small scale orography like valley winds, variations in near-surface temperature, orographically induced precipitation as well as river discharge for small and medium sized catchments cannot be modelled accurately or are even missing in simulations with 50 km resolution. For such purposes, a horizontal resolution below 20 km is necessary to resolve the high horizontal and time-dependent variability of meteorological variables. The few regional climate studies that have been carried out with horizontal resolutions finer than 20 km reveal that such high-resolutions are able to realistically capture regional climate features (CHRISTENSEN et al. 1998, LEUNG et al., 2004). However, the increase of resolution does not necessarily lead to a better performance (NOBRE et al., 2001). The error from global models is carried over into the regional models and often this error increases with increasing resolution (CHRISTENSEN et al., 1998) and depends on the variable considered. A detailed comparison between low and high-resolution runs is therefore necessary for each specific region to determine the added value of high-resolution simulations. Up to now, there have been many open questions regarding parameterisations and numerical issues in high-resolution regional climate simulations. Therefore, high-resolution climate simulations with horizontal resolutions finer than 20 km were performed and evaluated in this thesis to gain further knowledge about the added value of such simulations in complex topography and to study the sensitivity of the results with respect to parameterisations and numerical algorithms used. The investigation area for all high-resolution simulations, as presented here, is the region of Southwest Germany. Due to the geographical situation (vicinity of Alps) and the complex topography, the region is adequate to demonstrate the skill and to detect the problems of high-resolution climate simulations.

In regional climate simulations, the adequate simulation of the processes at the surface becomes more and more important due to the fact that the model is able to resolve mesoscale phenomena. Mesoscale phenomena, e.g. valley winds or convection, are highly influenced by the temperature of the soil and the soil moisture content. To improve the prediction of the surface parameters, which are the lower boundary condition of the atmospheric part of the regional climate model, and to study the influence of the soil-vegetation model, the soil-vegetation model VEG3D (SCHÄDLER, 1990) is implemented into COSMO-CLM in this thesis. In contrast to the operationally used soil-vegetation model TERRA_LM (DOMS et al., 2005), VEG3D has an explicit vegetation layer and should therefore yield better simulation results for high vegetation than TERRA_LM. The new scheme is expected to improve the modelling of soil processes and water cycle in COSMO-CLM.

An estimation of the consistency of the results obtained by regional climate simulations can only be obtained by performing ensemble simulations, because no estimate of the uncertainties can be given in the results with one single climate simulation. First projects of assessing and evaluating climate change for present and future from ensembles of regional climate model simulations and for assessing the uncertainties in such predictions have been carried out or are ongoing in the EU-projects PRUDENCE (CHRISTENSEN and CHRISTENSEN, 2007) and ENSEMBLES (HEWITT, 2005). Ensembles can be created by using different driving data sets, different regional climate models, different model parameterisations, different model setups (e.g. different horizontal resolution or parameter data sets), and perturbing the initial and boundary conditions. In this thesis, the uncertainty in the high-resolution simulations is determined by ensemble simulations using different driving data sets, model parameterisations, and different model setups.

Ensemble simulations for climate time scales are very time consuming and methods are therefore being developed that may replace the explicit simulation of every single day. One possibility is statistical-dynamical downscaling. This method was applied in this thesis and its potential in regional climate prediction was investigated.

Before using a regional climate model for the prediction of future climate, evaluations of the model results for the past and present are performed to assure that the model is able to reproduce the current climate right. This is often done for the time period from 1960 to 2000 due to the observation density that is highest during this time, and reliable driving data sets are available. As the climate change is most pronounced during the last two decades and simulations for a climatological 30-year period are too time consuming, the evaluation of the high-resolution simulations was done for the 1990s in this thesis. The focus of evaluation is on the 2m-temperature as an important meteorological variable to detect climate trends and on precipitation as one important variable of the hydrological cycle. The change in the water balance of the soil is investigated. Due to the lack of adequate observations this can only be done qualitatively.

The aim of this thesis is to gain deeper knowledge about the sensitivity of high-resolution regional climate simulation with COSMO-CLM in different model setups and also to improve the simulation results. This is done by (1) the performance and evaluation of ensemble simulations for the area of Southwest Germany, which are used to determine a suitable model setup for such simulations for that region, (2) the implementation of a new soil-vegetation model to determine the influence of the lower boundary on high-resolution regional climate simulations and (3) the introduction of a statistical-dynamical downscaling method to save computing time.

In the first part of this thesis ensemble simulations are evaluated to determine an optimal simulation set-up for high-resolution simulations in Southwest Germany and to determine the uncertainties in such simulations. The regional climate simulation results and the driving data from the global models were evaluated against observations to determine the error propagation from global to regional simulations. The evaluation of the high-resolution climate simulations was done using DWD observation data provided by PIK.

In the second part of this thesis, simulations with both soil-vegetation models in stand-alone mode and online coupled to COSMO-CLM are compared to determine the influence of soil-vegetation model on climate simulations. For climate simulations, freezing and melting processes in the soil had to be implemented into VEG3D in order to obtain realistic simulations of soil temperature and energy transfer at the surface during the cold seasons.

A third topic of this thesis is the development of a strategy for an efficient statistical–dynamical downscaling scheme to replace time consuming continuous climate simulations. This saving of time would enable the performance of a high amount of ensemble simulations to statistically evaluate climate change predictions.

The thesis is structured as follows. An introduction into regional climate modelling is given in the second chapter. In the third chapter, a description of the COSMO model is provided. Sensitivity studies for the determination of an adequate model setup for high-resolution regional climate simulations for Southwest Germany are presented in chapter four. In chapter five, the further development of VEG3D, the introduction of freezing and melting processes in the soil, and the coupling with COSMO-CLM are explained. In chapter six, comparisons between VEG3D and TERRA_LM in stand-alone mode and online coupled to COSMO-CLM are shown. The development of a statistical-dynamical downscaling method and comparisons with continuous simulations and observations are presented in chapter seven and conclusions and an outlook are given in chapter eight.

2 REGIONAL CLIMATE MODELS AND SIMULATIONS

The necessity of regional climate simulations for climate impact studies has been recognized since the end of the 1980s. At that time, general circulation models (GCM) had a resolution of about 300 - 500 km and it was therefore not possible to use GCM results for regional climate impact studies (HOUGHTON et al., 1990). The first regional climate simulations were carried out by DICKINSON et al. (1989) and GIORGI and BATES (1989). The term "regional" in this context means scales of 10^4 to 10^7 km^2 (IPCC, 2001). Scales larger than 10^7 km^2 are called "planetary scale" and scales lower than 10^4 km^2 are referred to as "local scale". The horizontal resolution of recent regional climate simulations reaches from about 50 km to 10 km.

Today, GCMs are run on scales lower than 50 km in operational weather forecast mode, e.g. GME (MAJEWSKI et al., 2002), and variable resolution GCMs are developed that can be run at any horizontal resolution. In the future, these variable GCMs may replace the RCMs in regional climate modelling. Nevertheless, the development of variable resolution GCMs is just in its beginnings and it is unlikely that GCMs will run on a 10 km scale for climate time scales in the nearer future (LEUNG et al., 2003). Besides the adaptation of physical parameterisations in GCMs (e.g. convection parameterisation) that would be necessary for using GCMs at different horizontal resolution, there is also the problem of limitation in computer power. Until computer power reaches the required level, regional climate simulations with limited area models are a growing field with more and more applications in the climate impact research.

Regional climate models (RCM) are limited area models and therefore need driving data at their lateral boundaries. Generally, GCM results are used as driving data and this process is called nesting a RCM into a GCM. Normally, one-way nesting is applied. This means that the RCM takes the GCM results as driving data at its lateral boundaries but the results obtained by the RCM are not fed back to the GCM simulation. The boundary relaxation of RCM results to the GCM results is often done in line with DAVIES (1976) with an exponentially decreasing weight in the buffer zone. This buffer zone consists of up to 10 grid boxes at each lateral boundary.

RCMs should not merely interpolate GCM results but they should be able to simulate local atmospheric feedback mechanisms that cannot be resolved by the coarse grid size of GCMs. This added value and the downscaling ability of nested RCMs has been ascertained by "big brother" experiments, which can help making decisions about domain size, location of simulation area, resolution jump from GCM to RCM and update frequency of lateral boundary data (DENIS et al., 2002, DENIS et al., 2003). In such experiments, first a reference RCM simulation (called the Big Brother) is performed over a large domain and then this data is used to drive another RCM (the Little Brother) on a smaller domain. The effect of degrading the resolution of lateral boundary conditions, spatially and temporally, is studied by comparing the big and little brother results.

The RCM domain size should be large enough that the model can simulate mesoscale effects without too much influence of the lateral boundaries (MCGREGOR, 1997) and it should be small enough that the circulation in the RCM does not differ too much from the circulation given by the GCM (JONES et al., 1995). Otherwise this difference in circulation patterns can lead to inconsistency errors at the outflow boundaries. One way to avoid such inconsistencies is to force the RCM to follow the GCM patterns by the spectral nudging technique (STORCH et al., 2000).

Forecasts of GCMs or reanalysis data sets can be used as driving data for the RCM. Reanalysis data sets are produced by running GCMs for past decades with assimilated observations. Therefore, they can be considered as the best method of interpolating observations to a regular grid. In general, reanalysis data is better suited to drive a RCM than forecast GCM data (ROJAS and SETH, 2003). The jump in horizontal resolution between GCM and RCM can be up to a ratio of 12 to yield realistic results for regional climate simulations of 45 km horizontal resolution (DENIS et al., 2003). For RCMs with a horizontal resolution of about 45 km, the lateral boundaries should be updated at least every 12 hours. Between simulations with an update every 3 and every 6 hours only small differences can be noticed (DENIS et al., 2003). The update period should be smaller than one quarter of the ratio of the length scale to the phase speed of the meteorological phenomena that penetrate over the boundaries into the RCM domain.

Most RCMs are hydrostatic (i.e. the equation for the vertical velocity is reduced to the hydrostatic equation) such as REMO (JACOB and PODZUN, 1997) or RegCM (GIORGI et al., 1993), but there are also non-hydrostatic RCMs such as MM5 (GRELL et al., 1994), CRCM

(CAYA and LAPRISE, 1999) and COSMO-CLM (BÖHM et al., 2006). Non-hydrostatic models represent deep convection and mountain waves with large vertical motion better, which is important for horizontal resolutions finer than 10 km. However, the added value by non-hydrostatic models has yet to be proven because most intercomparison projects focus on horizontal resolutions of 30 km and beyond. From operational weather forecasting it is well known, that for explicit simulation of convection, for example, the hydrostatic approximation does not hold.

As is the case of GCMs, a variety of parameterisations is used in regional climate models. The resulting spread of simulation results obtained by using different regional climate models has been investigated in many studies (MERCURE over Europe, PIRCS over the United States, RMIP over Asia, IRI/ARCs over South America (WANG et al., 2004 and references therein)). The large-scale patterns are well represented by nearly all RCMs but there are large differences between the single models, depending on the region and the season considered. A present-day climate model intercomparison on 50 km scale shows a warm bias of RCMs in winter and summer and a cold bias in transition seasons compared to observations (JACOB et al., 2007). In summer, a better agreement of observed and simulated values of inter-annual variability is found than in winter. The third IPCC Assessment Report found that RCMs with a horizontal resolution of 50 km have a bias of +-2°C and a precipitation bias of +/- 50% (IPCC, 2001). The model error in GCMs is about +/- 5°C and -40 to 60% in precipitation (LEUNG et al., 2003).

The predictive power of RCMs is mainly limited by three factors: the uncertainties in driving data, the uncertainties in physical parameterisations, and the chaotic nature of the climate system (WANG et al., 2004). RCMs show quite realistic climate signals when compared to observations, but are unable to "correct" systematic errors in large-scale circulation from driving data (CHRISTENSEN et al., 1998).

Furthermore, the results of RCMs can be influenced by vegetation parameters used for modelling the exchange between soil and atmosphere, which are provided as external parameters from different land use maps. The use of different parameter data sets can cause differences in monthly mean 2m-temperature of up to 1.5 K (BLOCK, 2007).

An often discussed topic is the evaluation of climate simulations. In the past this was primarily done by using gridded data sets (e.g. Climate Research Unit (CRU) data (NEW et al., 2000), WILLMOTT and MATSUURA (2001)). One reason for this, and a serious disadvantage, was the low density of the observations (IPCC, 2001). In the past decade, the density in observation stations has increased considerably. Hence, more recent studies also use station data to evaluate the results of the nearest grid box (MOHBERG and JONES, 2004). The grid box value of a simulation is always an area mean and a station value is a point observation that cannot be compared directly if the model resolution is too coarse. Even when high density observations are available we have to take into account the measurement error in our comparisons of climate simulations with observations. Especially for precipitation, the correction factor given by some studies is in the range of 15% (FREI et al., 2003) and for winter periods, where the differences are larger, even in the range of 30% (HAD, 2003).

Dynamical downscaling using RCMs is only one method for downscaling GCM results on regional scales. There are also statistical methods used for downscaling (WILBY et al., 2002). In dynamical downscaling, a high-resolution RCM is nested into a coarser resolution GCM and every day of the investigated period is simulated by a continuous simulation with the RCM (MCGREGOR, 1997). The advantage of this method is its ability to resolve small-scale atmospheric features, e.g. valley winds and orographic precipitation. The disadvantage is the high computational demand due to the explicit simulation of every day.

In statistical downscaling it is assumed that regional climate is mainly a function of the large-scale atmospheric state. Large-scale variables, known as predictors, from GCMs are used to determine the regional climate variables as predictands (IPCC, 2001). The assumption is made that the relationships found between them are also valid in the future. Three main techniques can be identified in statistical downscaling: (1) weather generators that randomly generate realistic looking sequences of local climate variables depending on GCM results; (2) transfer functions that give a quantitative relationship between predictor and predictand; (3) weather typing schemes that relate an atmospheric large-scale state to local climate variables.

A mixture of dynamical and statistical downscaling is the statistical-dynamical downscaling (FUENTES and HEIMANN, 2000, FREY-BUNESS et al., 1995). Typical weather patterns from GCM results are classified and each weather pattern is represented by its frequency. For each weather pattern, a RCM is used to simulate the according weather period and the resulting daily means of meteorological variables are summed up, weighted by their frequency, to receive annual mean values. The advantages of this method are that it is valid under future conditions because GCM forecasts can be used as driving and classification data and that it is not as time consuming as dynamical downscaling.

The growing interest of administrations and policymakers in regional climate change and climate impact studies has led to many studies of trend detection and future climate change for individual regions. The KLIWA project has been initiated for Southwest Germany, which aims to detect present and future climate changes with a focus on hydrology. It was found that the climate conditions have changed especially during the last three decades (GÜNTHER, 2004, STRAUB, 2004). Comparisons of the results of different downscaling techniques for present and future climate have been carried out within this project with the aim to provide reliable results of the influence of climate change on the water cycle. The result was that the statistical-dynamical method gives better results for present day climate than the dynamical downscaling for Southwest Germany (KLIWA, 2006). However, regional climate models have improved considerably and studies with the improved models are under way.

Summaries on regional climate modelling, downscaling techniques and further challenges in regional climate simulation can be found in MCGREGOR (1997), GIORGI and MEARNS (1999), IPCC (2001), LEUNG et al. (2003) and WANG et al. (2004).

The regional model COSMO-CLM that is used for the following studies is described in detail in Doms et al. (2005), Böhm et al. (2006) and Will et al. (2007). A summary of model history and a short model description is given in chapter three; in the following chapters only the components relevant for the related study and the parameterisations used are described.

Based on the information described above, a model setup, simulation chain and evaluation strategy has been composed for the following studies: reanalysis data is used as driving data for the regional climate simulations of past decades. A downscaling chain is performed to keep the jump in horizontal resolution within reasonable limits: first a coarser resolution COSMO-CLM simulation is nested into the reanalysis data with an update frequency of boundary data of six hours and then the results of the coarse regional model are used to drive the high-resolution simulations with an update frequency of three hours (Fig. 2.1). The domain size for the regional model simulations is chosen small enough to ensure there are no large inconsistencies between RCM and GCM (WILL, personal communication, 2005). The horizontal resolution used for the high-resolution runs is well below 20 km. The validation of the high-resolution climate simulations has been done using observation data obtained from the PIK (ÖSTERLE et al., 2006). In this thesis, the evaluation of climate simulations is done by comparing point measurements with the weighted mean value of the nearest four grid boxes of the simulations.

In the following studies mainly the results of simulations for 2m-temperature and precipitation are discussed. One should keep in mind that these are only the results of a long process chain and that 2m-temperature is just a diagnostic variable. Therefore, it would be more suitable for model validation to evaluate variables like heat fluxes, integrated moisture content for the grid box column, or radiation balance components. However, area-wide observations of these parameters are not yet available. Hence, it is generally assumed that when the regional model is able to reproduce precipitation and 2m-temperature, all other processes like circulation patterns and three dimensional meteorological variables are reasonably well simulated.

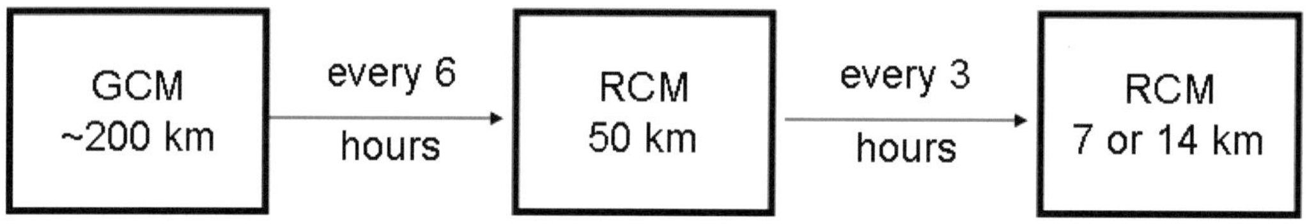

Figure 2.1: Downscaling chain for high-resolution runs.

3 DESCRIPTION OF THE COSMO MODEL

The Lokal-Model (LM) has been the operational limited area forecast model of the DWD since December 1999. LM is designed for meso-β and meso-γ scales (Fig. 3.1). Especially on meso-γ scales non-hydrostatic effects begin to play an essential role (DOMS et al., 2005). Therefore, the LM equations are non-hydrostatic and fully compressible. More than 10 European meteorological services are participating in COSMO (Consortium for Small-Scale Modelling) and are running and developing the LM for operational weather forecast. In 2007, LM was renamed the COSMO model to reflect the joint effort. The model is available free of charge for research purposes and the mode in which the model is run is specified by the appendix to the model name. For example, COSMO-EU is the operational forecast mode used at DWD with a horizontal grid spacing of 7 km, a domain covering the whole of Europe and data assimilation. COSMO-DE is the short range weather prediction mode at DWD running on a 2.8 km horizontal resolution and an area covering the whole of Germany. COSMO-CLM is the climate mode of the model with no prescribed horizontal resolution or simulation area, with updated vegetation parameters and with no data assimilation. All descriptions of dynamics, numerics and parameterisations given in the following chapter are valid for the forecast and the climate mode of the COSMO model.

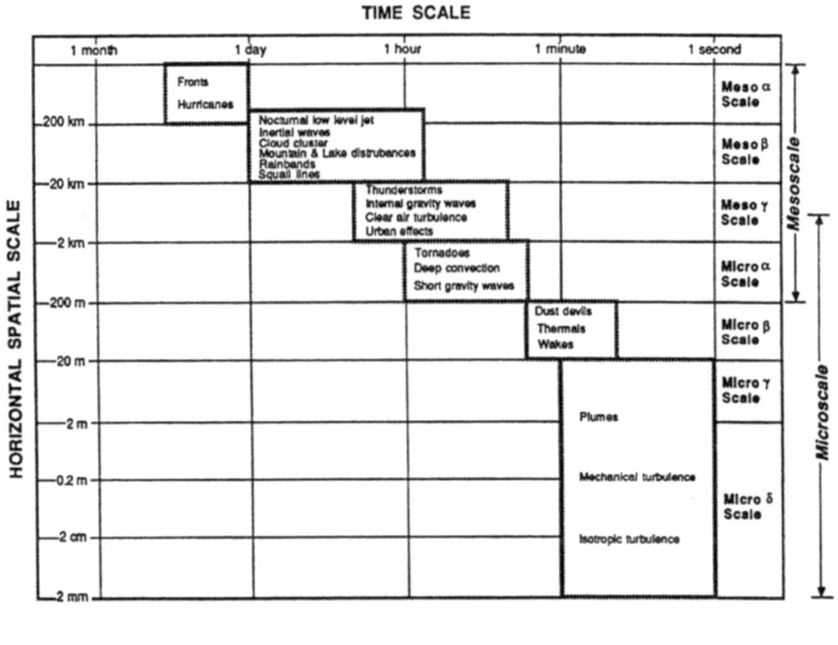

Figure 3.1: Definition of meso- and microscale (STULL, 1988).

3.1 Dynamics

3.1.1 Model equations

The equations in this subsection have been taken over from the model description of Doms and Schättler (2002) (page 9 and 48/49). The COSMO model uses the basic conservation laws for momentum, mass and heat:

$$\rho \frac{d\mathbf{v}}{dt} = -\nabla p + \rho \mathbf{g} - 2\mathbf{\Omega} \times (\rho \mathbf{v}) - \nabla \cdot \underline{\mathbf{t}}, \tag{3.1}$$

$$\frac{d\rho}{dt} = -\rho \nabla \cdot \mathbf{v}, \tag{3.2}$$

$$\rho \frac{dq^x}{dt} = -\nabla \cdot \mathbf{J}^x + I^x, \tag{3.3}$$

$$\rho \frac{de}{dt} = -p \nabla \cdot \mathbf{v} - \nabla \cdot (\mathbf{J}_e + \mathbf{R}) + \varepsilon . \tag{3.4}$$

Bold symbols are used to represent vectors and bold underlined symbols indicate dyadic tensors. Scalar and vector products are indicated by · and × respectively.

The following symbols are used:

t: time

p: pressure

T: temperature

ρ^x: partial density of mixture constituent x

ρ: total density of air mixture

q^x: mass fraction (specific content) of constituent x

e: specific internal energy

$\mathbf{v}$: barycentric velocity (relative to the rotating earth)

$\mathbf{I}^x$: sources/sinks of constituent x

$\mathbf{J}_e$: diffusion flux of internal energy

$\mathbf{J}^x$: diffusion flux of constituent x

$\mathbf{R}$: flux density of solar and thermal radiation

$\underline{\mathbf{t}}$: stress tensor due to viscosity

ε: kinetic energy dissipation due to viscosity

$\mathbf{g}$: apparent acceleration of gravity

$\mathbf{\Omega}$: constant angular velocity of earth rotation

∇: gradient (Nabla) operator

$\frac{d}{dt}$: total (Lagrangian) time derivative operator

The index x represents a specific constituent of the mixture with

x = d	for dry air,
x = v	for water vapour,
x = l	for liquid water, and
x = f	for water in the solid state, i.e. ice.

From this set of basic equations written in the advection form an alternative set of basic equations can be derived, consisting of prognostic equations for temperature, pressure, wind velocity, and mixing ratios and a diagnostic equation for the air density. The meteorological variables are averaged over the volume of the grid box and over the time step. Afterwards, they are split up into a mean value (the grid scale value) and a deviation from this mean value (the sub-grid scale value). The sub-grid scale values are then calculated by physical parameterisations. Some simplifications are made before these basic equations are transformed to a rotated spherical grid: (1) all molecular fluxes are neglected except the diffusion fluxes of the liquid and solid forms of water, (2) the specific heat of moist air is replaced by the specific heat of dry air and (3) the pressure change due to changes in the concentrations of the water constituents resulting from diffusion fluxes and phase transitions are neglected. Afterwards, a hydrostatic balanced base state at rest is introduced and all model variables are described as a sum of base state and grid-scale deviation (perturbation). This has the advantage of removing horizontal base-state pressure gradient terms in the equation of motion and this reduces the computational error in the calculation of the pressure gradient force in the case of sloping coordinate surfaces. The equations are then transformed to a time-independent terrain-following coordinate system to produce a simplified formulation of the lower boundary conditions. Operationally, a pressure-based hybrid coordinate is used. A detailed description of the transformation steps can be found in DOMS and SCHÄTTLER (2002). Eventually, seven prognostic equations and one diagnostic equation for horizontal wind velocity, vertical wind velocity, perturbation pressure, temperature, water vapour, liquid and solid forms of water, and total density of air form the basis of the model system:

- Horizontal wind velocity:

$$\frac{\partial u}{\partial t} = -\left\{ \frac{1}{a\cos\varphi}\frac{\partial E_h}{\partial \lambda} - vV_a \right\} - \dot{\zeta}\frac{\partial u}{\partial \zeta} - \frac{1}{\rho a\cos\varphi}\left(\frac{\partial p'}{\partial \lambda} - \frac{1}{\sqrt{\gamma}}\frac{\partial p_0}{\partial \lambda}\frac{\partial p'}{\partial \zeta} \right) + M_u \tag{3.5}$$

$$\frac{\partial v}{\partial t} = -\left\{ \frac{1}{a}\frac{\partial E_h}{\partial \varphi} - uV_a \right\} - \dot{\zeta}\frac{\partial v}{\partial \zeta} - \frac{1}{\rho a}\left(\frac{\partial p'}{\partial \varphi} - \frac{1}{\sqrt{\gamma}}\frac{\partial p_0}{\partial \varphi}\frac{\partial p'}{\partial \zeta} \right) + M_v \tag{3.6}$$

- Vertical wind velocity

$$\begin{aligned} \frac{\partial w}{\partial t} = & -\left\{ \frac{1}{a\cos\varphi}\left(u\frac{\partial w}{\partial \lambda} + v\cos\varphi\frac{\partial w}{\partial \varphi} \right) \right\} - \dot{\zeta}\frac{\partial w}{\partial \zeta} + \frac{g}{\sqrt{\gamma}}\frac{p_0}{\rho}\frac{\partial p'}{\partial \zeta} + M_w \\ & + g\frac{\rho_0}{\rho}\left\{ \frac{(T - T_0)}{T} - \frac{T_0 p'}{T p_0} + \left(\frac{R_v}{R_d} - 1 \right) q^v - q^l - q^f \right\} \end{aligned} \tag{3.7}$$

- Perturbation pressure

$$\frac{\partial p'}{\partial t} = -\left\{ \frac{1}{a\cos\varphi}\left(u\frac{\partial p'}{\partial \lambda} + v\cos\varphi\frac{\partial p'}{\partial \varphi} \right) \right\} - \dot{\zeta}\frac{\partial p'}{\partial \zeta} + g\rho_0 w - \frac{c_{pd}}{c_{vd}} pD \tag{3.8}$$

- Temperature

$$\frac{\partial T}{\partial t} = -\left\{ \frac{1}{a\cos\varphi}\left(u\frac{\partial T}{\partial \lambda} + v\cos\varphi\frac{\partial T}{\partial \varphi} \right) \right\} - \dot{\zeta}\frac{\partial T}{\partial \zeta} - \frac{1}{\rho c_{vd}} pD + Q_T \tag{3.9}$$

- Water vapour

$$\frac{\partial q^v}{\partial t} = -\left\{ \frac{1}{a\cos\varphi}\left(u\frac{\partial q^v}{\partial \lambda} + v\cos\varphi\frac{\partial q^v}{\partial \varphi} \right) \right\} - \dot{\zeta}\frac{\partial q^v}{\partial \zeta} - \left(S^l + S^f \right) + M_{q^v} \tag{3.10}$$

- Liquid and solid forms of water

$$\begin{aligned} \frac{\partial q^{l,f}}{\partial t} = & -\left\{ \frac{1}{a\cos\varphi}\left(u\frac{\partial q^{l,f}}{\partial \lambda} + v\cos\varphi\frac{\partial q^{l,f}}{\partial \varphi} \right) \right\} - \dot{\zeta}\frac{\partial q^{l,f}}{\partial \zeta} - \frac{g}{\sqrt{\gamma}}\frac{\rho_0}{\rho}\frac{\partial P_{l,f}}{\partial \zeta} \\ & + S^{l,f} + M_{q^{l,f}} \end{aligned} \tag{3.11}$$

- Total density of air

$$\rho = p\left\{ R_d \left(1 + \left(\frac{R_v}{R_d} - 1 \right) q^v - q^l - q^f \right) T \right\}^{-1} \tag{3.12}$$

The symbols in equations 3.5 to 3.12 have the following meaning:

T_0, p_0, ρ_0 : base state values of temperature, pressure and density

S^l, S^f : cloud microphysical sources and sinks due to phase change

M_u, M_v, M_w, M_{q^x} : source terms due to small scale turbulent mixing, subgrid scale moist convection, lateral boundary relaxation, computational mixing, and Rayleigh damping scheme

c_{vd}: : specific heat of water vapour at constant pressure

c_{pd} : specific heat of dry air at constant pressure

D: divergence of the wind field

Q_T: diabatic heating

$P_{l,f}$: precipitation fluxes

R_v : gas constant for water vapour

R_d : gas constant for dry air

φ : rotated latitude

ρ: density

λ: rotated longitude

ζ: terrain following vertical coordinate

p': grid scale pressure deviation

$\dot{\zeta}$: contravariant vertical velocity

$\sqrt{\gamma}$: variation of reference pressure with ζ

E_h : kinetic energy of horizontal motion

V_a : absolute vorticity

3.1.2 ROTATED COORDINATE SYSTEM

Typical applications of the model cover areas of the size of a few million square kilometres (e.g. the whole of Europe), which makes it necessary to take the curvature of the Earth into account. Therefore, the model equations are written in spherical coordinates. However, when using spherical coordinates, two problems arise. One is the "pole problem", which means that the geographical poles represent a singularity due to the convergence of the meridians, and therefore special measures have to be taken when a geographical pole lies within the simulation domain. More often, however, the second problem (also connected with the convergence of the meridians) is encountered: varying horizontal resolution with latitude away from the equator. A suitable way to avoid both problems is realised in

COSMO: the rotated grid. The computational spherical coordinate system is rotated in such a way that the intersection of the equator and the prime meridian of the new system passes through the simulation domain, thus avoiding the pole problem and providing minimal convergence of the meridians at the same time (Fig. 3.2). The necessary coordinate transformations are performed during pre and postprocessing.

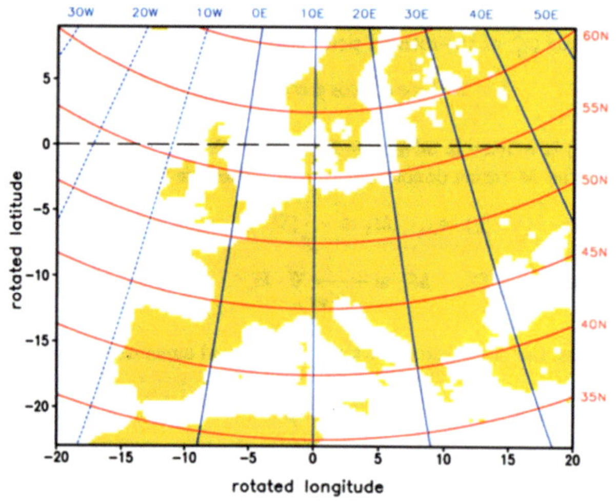

Figure 3.2: Geographical longitude (blue) and latitude (red) in the unrotated grid. The dashed line indicates the equator in the rotated grid with pole coordinates 32.5°S and 10.0°E in the unrotated system. The rotated 0° meridian corresponds to the 10°E geographical meridian (DOMS and SCHÄTTLER, 2002).

3.1.3 TERRAIN FOLLOWING COORDINATES

In the vertical, a time-independent terrain-following coordinate system is used with user-defined grid stretching. The vertical coordinate is specified by a unique transformation relation $z = f(\lambda, \varphi, \zeta)$. A twostep procedure is applied to keep the numerical formulation of the model equations independent of the choice of ζ: first a terrain-following transformation is done using a user-defined coordinate $\tilde{\zeta}$ and then $\tilde{\zeta}$ is mapped to the computational coordinate ζ using a monotonic function m in the form $\tilde{\zeta} = m(\zeta)$ (Fig. 3.3). Three options for the terrain-following coordinate $\tilde{\zeta}$ are available: (1) a reference pressure based coordinate, (2) a Gal-Chen height-based coordinate, and (3) a height-based SLEVE (Smooth Level Vertical) coordinate.

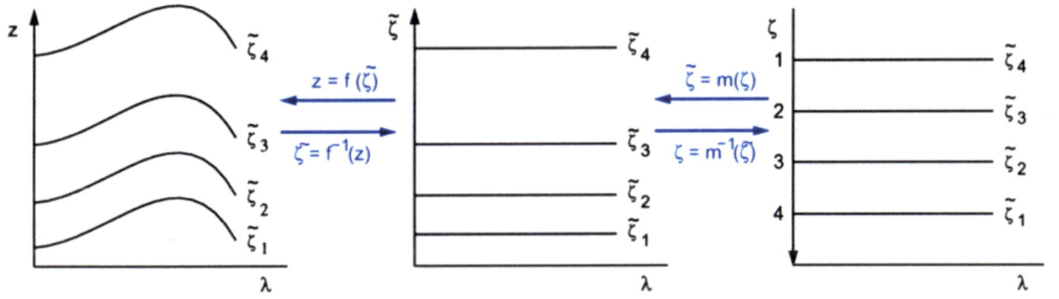

Figure 3.3: Mapping of an irregular curvilinear grid with a terrain following coordinate $\tilde{\zeta}$ onto a rectangular equidistant grid ζ (DOMS and SCHÄTTLER, 2002).

3.2 NUMERICS

3.2.1 GRID STRUCTURE

The meteorological variables are staggered on an Arakawa C-grid horizontally. Vertically, a Lorenz grid staggering is applied: the scalars are defined at the centre of a grid box (main level) and the normal velocity components are defined on the corresponding box faces (half level in the vertical) (Fig. 3.4, Fig. 3.5).

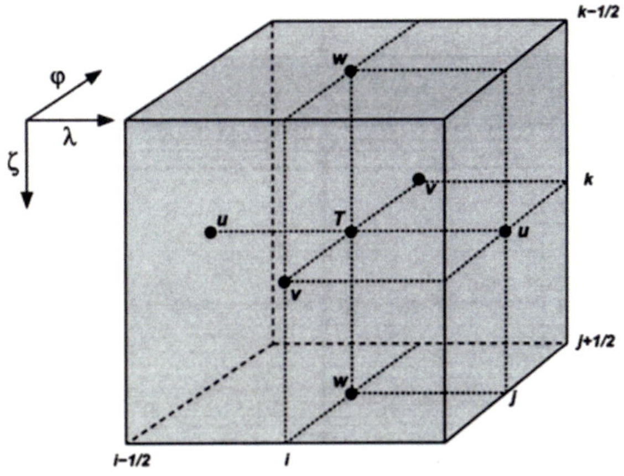

Figure 3.4: A grid box volume showing the Arakawa-C/Lorenz staggering of the prognostic dependent model variables (DOMS and SCHÄTTLER, 2002). T represents all scalar variables.

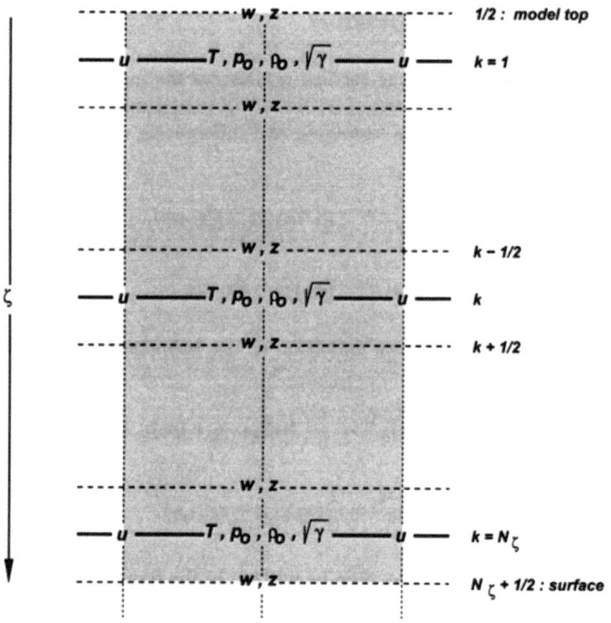

Figure 3.5: Vertical staggering of variables and metric terms in a grid box column with N_ζ layers (DOMS and SCHÄTTLER, 2002). Dashed lines are the model half levels separating the main levels (full lines).

3.2.2 TIME INTEGRATION

Numerically, the prognostic equations are solved by a three time-level leapfrog method (horizontally explicit, vertical implicit) for horizontal grid size resolutions coarser than 3 km. A time-split integration scheme, including extensions proposed by SKAMAROCK and KLEMP (1992) is used in the leapfrog scheme: the prognostic equations are separated in terms containing fast and slow processes. This operator splitting is necessary because the unfiltered equations contain sound and gravity waves, which normally require a rather small time step. In the leapfrog scheme, terms containing slow processes (e.g. advection) are stepped forward from the time level $t-\Delta t$ to time level $t+\Delta t$ and terms containing fast processes are stepped forward from time level $t-\Delta t$ to time level $t+\Delta t$ in n smaller steps Δt_S. This splitting makes the code much more efficient (Fig. 3.6).

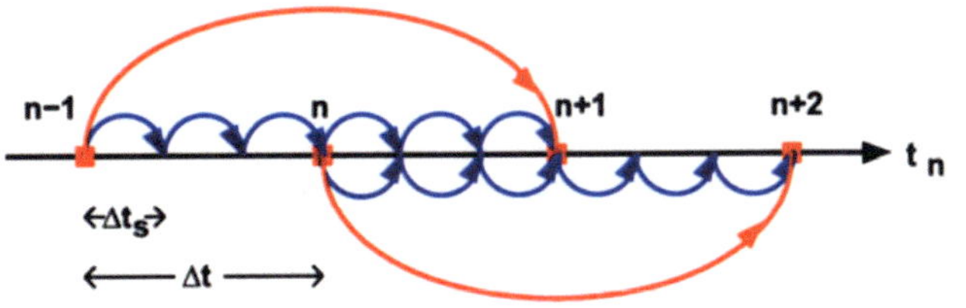

Figure 3.6: The KLEMP and WILHELMSON (1978) time-splitting algorithm (DOMS and SCHÄTTLER, 2002).

3.3 INITIAL AND BOUNDARY CONDITIONS

For COSMO-CLM, five different data sets are available as driving data (Tab. 3.1). Additionally, COSMO simulations at high resolution can be driven by COSMO simulations with coarser horizontal resolutions. At the lateral boundaries, a one way nesting with Davies-type lateral boundary formulation is applied and at the top of the model domain a Rayleigh damping layer is used. The model has a sponge zone of several grid boxes where the simulations are adapted towards the driving data. The horizontal extent of the computational domain is chosen smaller than the total model domain to implement boundary conditions and to apply the domain decomposition strategy. In general, the boundaries of the physical domain within which the model equations are integrated numerically, are placed two grid points away from the border of the total domain. In the two grid point frame outside the physical domain all model variables are set to specific boundary values but no computation is done. After the initialisation, where the meteorological variables are taken over from the driving data set for the whole model domain, the lateral boundary conditions are updated regularly. The soil moisture and soil temperature profiles are only initialised once at the start of the simulation for the whole domain and are only updated at the lateral boundaries therafter.

Name	Approx. resolution (km)	Time covered	Type	Suitable for prognoses of future climate
ERA-40	125	1950 - 2002	Reanalyses	No
NCEP	210	1948 - present	Reanalyses	No
ECHAM5	150	1960 - 2100	Climate runs	Yes
GME	60	1999 - present	Analyses	No
HadCM3	300	1960 - 2100	Climate runs	Yes

Table 3.1: Available large-scale driving data sets for COSMO-CLM. ECHAM5 is the global model operated by the Max-Planck-Institute for Meteorology, Hamburg, GME is DWD's global weather forecast model and HadCM3 is the Hadley Centre Coupled Model version 3.

3.4 PHYSICAL PARAMETERISATIONS

Subgrid-Scale Turbulence

A prognostic turbulent kinetic energy closure at level 2.5 after MELLOR and YAMADA (1974) is applied, including effects from subgrid-scale condensation and from thermal circulations.

Surface Layer Parameterization

Surface scheme based on turbulent kinetic energy is applied, including a laminar-turbulent roughness layer.

Grid Scale Clouds and Precipitation

Bulk microphysics parameterisation for precipitation formation including four hydrometeor species (cloud droplets, raindrops, cloud ice and snowflakes). Three-dimensional transport of precipitation and prognostic precipitation is accounted for.

Subgrid-Scale Clouds

Subgrid-scale cloudiness is calculated by an empirical function, depending on relative humidity and height.

Moist convection

Moist convection is parameterised with the TIEDTKE (1989) mass-flux convection scheme, with equilibrium closure based on moisture convergence. Optionally, a KAIN-FRITSCH (1992) mass-flux convection scheme with non-equilibrium CAPE-type closure can be used. The main differences between the two schemes, besides the closure, are the differences in the triggering criteria for convection and the processes influencing detrainment and entrainment (SMOYDZIN, 2004).

Radiation

Radiation is parameterized according to a δ two-stream radiation scheme after RITTER and GELEYN (1992) for short and longwave fluxes. The parameterisation uses eight spectral intervals with full cloud-radiation feedback.

Soil model

The soil processes are parameterised by the soil-vegetation model TERRA_LM using prognostic equations for soil temperature and soil water content for several soil layers. A detailed description can be found in subsection 6.1.1.

Terrain and surface data

The model has its own databases for external data, like orography, land-sea mask, soil type and vegetation cover. The data sets are available for different horizontal resolutions and pre-defined regions covering Europe. New data sets can be created by a pre-processor program using high-resolution global data sets.

4 SENSITIVITY STUDIES

A prerequisite for the study of future changes is to asses how well and with what uncertainties the past and present climate, and particularly components of the regional water cycle like precipitation and evapotranspiration, can be modelled in orographically structured regions like Southwest Germany and to define an adequate simulation setup. Model results are influenced strongly by model setup. Changing the parameter settings can affect the quality of the results considerably. The most important parameters depend in general on the meteorological variables considered, so that no single set of parameters will give optimal overall results. The following parameters seem to have essential impact on the results and are therefore studied in this thesis:

Domain size: Every change in the size of the simulation area causes a change of simulation results and there is no recipe for an optimum domain size. Several studies have been carried out on this issue and they mainly draw three findings that should be considered when determining the domain size.

(a) The domain size should be chosen sufficiently small so that the synoptic circulation does not differ significantly from that of the driving global model (JONES et al., 1995).

(b) The domain size has to be chosen sufficiently large so that the simulation is not tied too strongly to the driving data if studies of internal forcing are carried out (SETH and GIORGI, 1998). WARNER et al. (1997) suggested placing the boundaries at a distance of at least half the size of the investigation area.

(c) Lateral boundaries should not be placed in regions with strong orographic forcing (DICKINSON et al., 1988). To save computing time, however, the domain should be as small as possible with still stable results. Before performing decadal simulations, it must be ascertained that no unphysical processes are induced by the choice of the model domain and that small changes of the domain size and position do not induce major changes in the simulation results.

Horizontal resolution: Up to now, regional climate models have been used commonly with resolutions of about 50 km for climate studies (JACOB et al., 2007, HAGEMANN et al., 2004). For a reliable representation of local effects, e.g. in hydrologically relevant fields, a much higher resolution is needed (CHRISTENSEN et al., 1998). It is not clear, however, which horizontal resolution is adequate to account for the small scale processes influencing, e.g. convective precipitation in orographically structured regions. Simulations with two different horizontal resolutions below 20 km will be compared to find out if a higher horizontal resolution (which costs an additional amount of computer resources) really yields better simulation results.

Driving data: Reanalysis data has been used as driving data for the simulation of past decades. Different numerical models, methods and observations create differences in the reanalyses (ZHAO et al., 2006, SIMMONS et al., 2004). NOUGER et al. (1998) found that in regional climate models, errors in summer precipitation and temperature are mainly forced by regional processes but that in winter the boundary forcing is responsible for about 40% of the total error variance for precipitation. Therefore, simulations for two different reanalysis data sets were performed to assess the variability in the simulations induced by using different driving data.

Model setup: A set of six sensitivity studies is conducted in order to quantify the uncertainties in simulations due to the use of different parameters, physical parameterisations, dynamical methods and initialisations. Within the COSMO model system two different convection schemes and two time integration schemes, as well as two different formulations for the CO_2 content for past and present climate, are available for use in this study. The number of hydrologically active layers is variable in the model and its influence will be investigated here. Furthermore the influence of soil moisture initialisation, which has been shown to have considerable influence in former studies (SETH and GIORGI, 1998, SCHÄR et al., 1999), will be assessed.

The simulation setup and the simulation domains are presented in the first section and in the second section, the impact of driving data, grid size (50 km, 14 km and 7 km), and soil initialisation on the simulation results is discussed for the whole simulation area. The evaluation of the model setup, concentrating on the area of Southwest Germany, is undertaken in the third section. A sensitivity study for the four parameters mentioned above is also presented. The most suitable model setup is then used to investigate the ability of COSMO-CLM to reproduce observed trends of the near-surface temperature between 1991 and 2000 for Southwest Germany.

4.1 SIMULATION SETUP

The simulations presented here were driven by reanalyses. This has the advantage that this data set is close to reality, making it possible to assess the capability of the regional model to simulate the climate correctly when driven with adequate data. As driving data for the simulations, ERA-40 (SIMMONS and GIBSON, 2000) and NCEP-DOE AMIP-II (KANAMITSU et al., 2002) reanalysis data sets were used. The coarse grid of the driving data makes an intermediate nesting step necessary. Therefore, the whole European region was simulated with a horizontal resolution of 0.44° (60x81 grid points) (Fig. 4.1) and the higher resolution simulations were nested therein (180x158 resp. 90x76 grid points for the 7 km and 14 km run). The area of the high-resolution runs will be referred to as alps domain. The 0.44° simulations (~ 50 km) were run from the first of June 1988 until the end of 2001. As initialisation values for the soil model in the 50 km simulations, climatological values provided by the ETH Zürich were used. This climatology data was obtained by averaging 30 years of COSMO-CLM simulation (Version 2.4.6) with 0.44° horizontal resolution driven by ERA-40 data. Details about the simulation can be found in JÄGER et al. (2007).

Southwest Germany is an orographically structured region bordered by the Vosges Mountains in the west and the Alps in the south; it includes the Black Forest, which reaches a maximum altitude of 1400 m, and the Swabian Jura, with a maximum altitude of 1000 m. The area of investigation includes the Federal State of Baden-Württemberg, the western part of Bavaria and the southern parts of Hesse (Fig. 4.1). For the evaluations in subsection 4.3.1, gridded precipitation data sets obtained with the method described in FRÜH et al. (2007) were used. Observations from 23 climate stations operated by DWD (ÖSTERLE et al., 2006) (Fig. 4.1 and Tab. 4.1) were used for the comparisons presented in subsection 4.3.2 to subsection 4.3.4. Tests have shown that the use of gridded data sets as well as the use of station data for comparison with the model results yields quite similar results.

For all high-resolution simulations, the model version 4.0 of COSMO-CLM was used. Only the 50 km simulations were run with the older model version 3.22. The physics in both versions used here is the same and they differ mainly in technical details. The time integration is realised by a three time level leapfrog scheme. The model physics contains a cloud scheme (DOMS et al., 2005), with prognostic precipitation and four hydrometeor species (cloud droplets, raindrops, cloud ice and snowflakes), a mass flux scheme for convection (TIEDTKE, 1989) and a delta-two-stream radiation scheme after RITTER and GELEYN (1992) called every hour. Turbulence is modelled by a turbulence and surface layer scheme using a prognostic turbulent kinetic energy equation on the basis of the level 2.5 scheme by

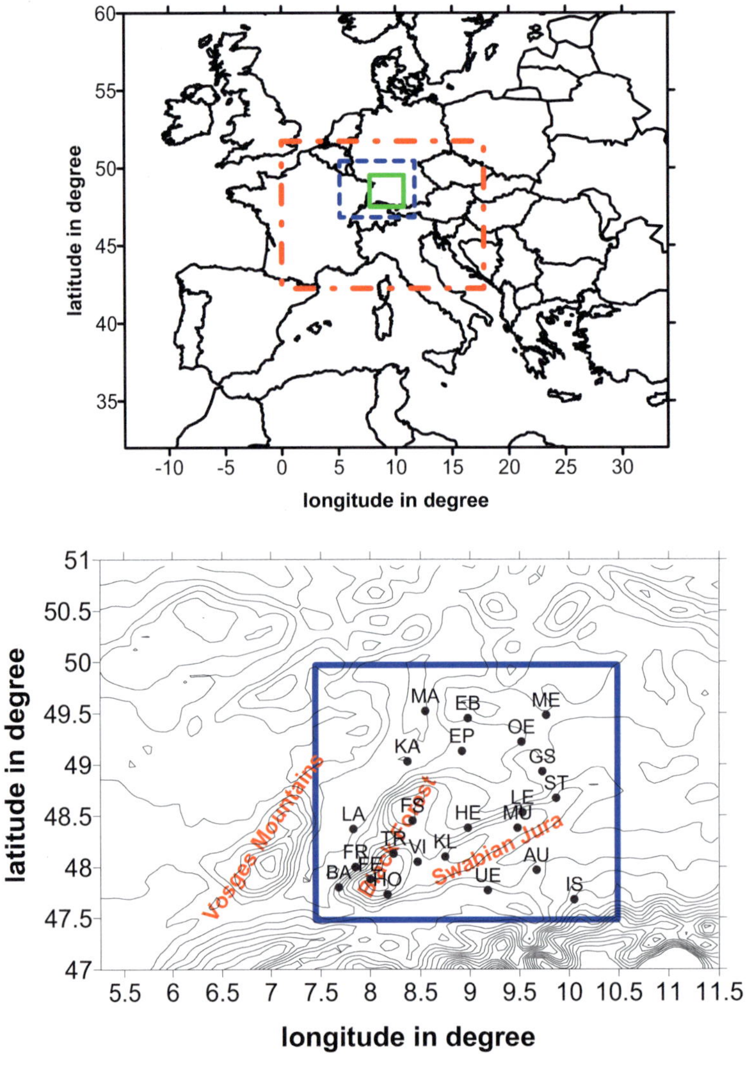

Figure 4.1: Top: Simulation domain for the 50 km runs. Red dash-dotted rectangle: Alps domain for the 7 km run. Blue dashed rectangle: Noalps domain for the 7 km run. Green solid rectangle: Investigation area. Bottom: Noalps domain with measurement stations. Evaluations are made inside the blue rectangle.

MELLOR and YAMADA (1974). The model contains the multi layer soil model TERRA_LM with prognostic soil temperature and water content (HEISE et al., 2003). The soil model comprises ten soil layers with a maximum depth of 15 meters. The main levels have the following depths (m): 0.005, 0.025, 0.07, 0.16, 0.34, 0.7, 1.42, 2.86, 6.74, 11.5. The external parameters, like vegetation parameters, topography and land-sea mask are taken from the data set provided by DWD.

Longitude (°)	Latitude (°)	Altitude	Abbreviation	Station name
9.67	47.97	560	AU	AULENDORF-SPIEGLER
7.68	47.8	367	BA	BADENWEILER
8.98	49.45	178	EB	EBERBACH/NECKAR
8.92	49.13	210	EP	EPPINGEN
8.00	47.88	1486	FE	FELDBERG/SCHW.
7.85	48.00	269	FR	FREIBURG I. BR. HBF
8.42	48.45	797	FS	FREUDENSTADT
9.73	48.93	492	GS	GSCHWEND KR. OSTALB
8.98	48.38	520	HE	HECHINGEN
8.17	47.73	1008	HO	HOECHENSCHWAND
10.05	47.68	712	IS	ISNY
8.37	49.03	112	KA	KARLSRUHE
8.75	48.10	973	KL	KLIPPENECK
7.83	48.37	155	LA	LAHR
9.53	48.53	758	LE	LENNINGEN-SCHOPFLOCH
8.55	49.52	96	MA	MANNHEIM
9.77	49.48	250	ME	MERGENTHEIM
9.48	48.38	750	MU	MUENSINGEN
9.52	49.22	276	OE	OEHRINGEN
9.87	48.67	734	ST	STOETTEN
8.23	48.13	683	TR	TRIBERG
9.18	47.77	490	UE	UEBERLINGEN/BODENSEE
8.47	48.05	720	VI	VILLINGEN - S.

Table 4.1: Coordinates, altitude, and abbreviation for the observation sites available for comparison with simulations.

4.2 SENSITIVITY STUDIES FOR THE WHOLE SIMULATION AREA

4.2.1 INFLUENCE OF DRIVING DATA AND HORIZONTAL RESOLUTION

Using the double nesting technique described above, there are three possible sources of errors in the high-resolution simulations for the alps domain:

1. The original reanalysis data sets can differ from the observational data and can cause spurious results of the 50 km simulations
2. The COSMO-CLM simulation with 50 km can produce unrealistic results even if the driving reanalysis data is correct and therefore produces spurious driving data for the alps domain
3. The 50 km simulations give realistic driving data for the alps domain, but the high-resolution simulations give spurious results.

Due to the lack of adequate observation data for an evaluation of the three dimensional meteorological variables only a comparison of near-surface temperature (this is defined as the temperature at 2 m height) and precipitation is presented here. The comparison is restricted to the land areas and the reanalysis data. The 50 km simulations are compared to the data set from WILLMOTT and MATSUURA version 1.02 (2001); this data set has a resolution of 0.5°. The comparison is made for the annual mean of near-surface temperature and the annual sum of precipitation average over the period from 1990 to 1999.
In the alps, domain differences between the two reanalysis data sets of up to 2 K and 300 mm occur for the annual mean near-surface temperature and the annual mean precipitation amount (Fig. 4.2, Fig. 4.3). The NCEP reanalysis data is warmer in the northern and the western part of the alps domain than the ERA data, and colder in the southern and eastern part. Besides the Alpine region and Southern France, the NCEP reanalysis data shows higher precipitation amounts than the ERA data. A comparison with observations shows differences of up to 4 K and 1000 mm in the alps domain (Fig. 4.2, Fig. 4.3). The reanalysis data sets are both colder and drier than the observations. One should keep in mind that data sets with different horizontal resolutions are compared. Therefore, some differences in temperature and precipitation can be explained by different terrain heights. Nevertheless, it

can be stated that there are systematic differences between the reanalysis data sets, which can lead to different simulation results in the 50 km simulations.

As for the second source of errors mentioned above, the temperature and precipitation differences between the simulations with 50 km resolution driven by ERA and NCEP data and the observations were calculated (Fig. 4.4, Fig. 4.5). Large negative temperature differences occur over major mountain ranges like the Alps. This may be due to the differences in terrain height between the real and the model topographies. For the rest of the alps domain, the differences between simulation and observations is lower than 1 K. Comparing the 50 km simulation results driven by ERA and NCEP (Fig. 4.4), one can see that the temperature difference is lower than 0.5 K in the alps domain. That means that the differences between the two 50 km simulations are lower for the alps domain than the differences between their driving data sets. This can be explained by the orography, which is the same in the two 50 km runs but differ between the two reanalysis data sets. Precipitation is overestimated by both models in the north of Europe, especially over the Alps and underestimated in the south of Europe (Fig. 4.5). As observed in other studies precipitation is overestimated at the northern site of the Alps and underestimated at the southern site (JÄGER et al., 2007). In the NCEP simulation precipitation is higher in the north of Europe and lower in the south as in the ERA simulation. In the alps domain the difference in annual precipitation amount between the two simulations is lower than 200 mm and therefore smaller than that of the reanalysis data sets.

The ERA simulation is colder and produces less rain in the alps domain than the simulation with NCEP data. The differences are also observed at the lateral boundaries of the alps domain and are therefore expected to be reflected by the high-resolution runs with 7 and 14 km. Over the year, westerly winds prevail. Therefore, we would expect higher temperatures in the 7 km run driven by the 50 km NCEP run due to the transport of warmer air from the boundaries into the simulation domain.

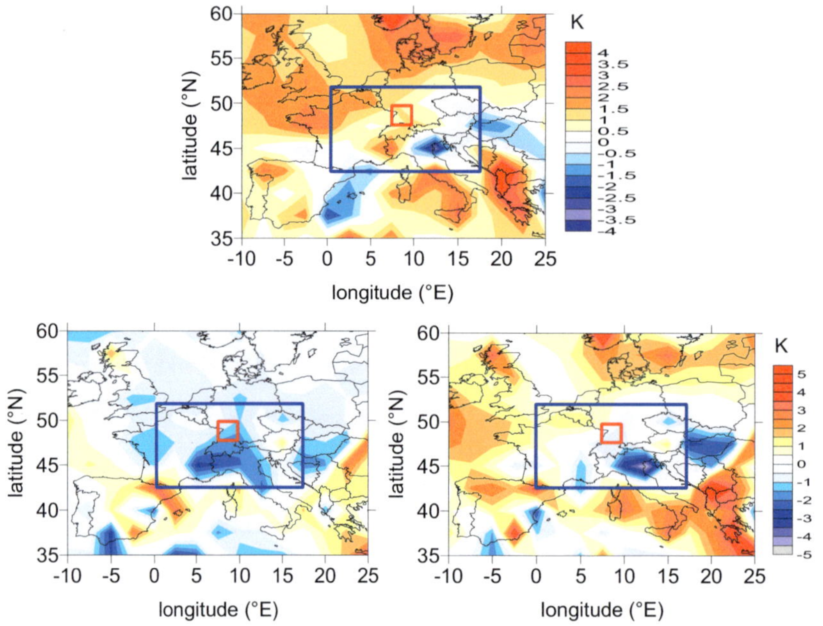

Figure 4.2: Annual mean near-surface temperature difference averaged over the period 1990 – 1999: difference between NCEP and ERA reanalysis (top), ERA and observations (bottom, left) and NCEP and observations (bottom, right). The red rectangle indicates the area of investigation and the blue rectangle the area of the alps domain.

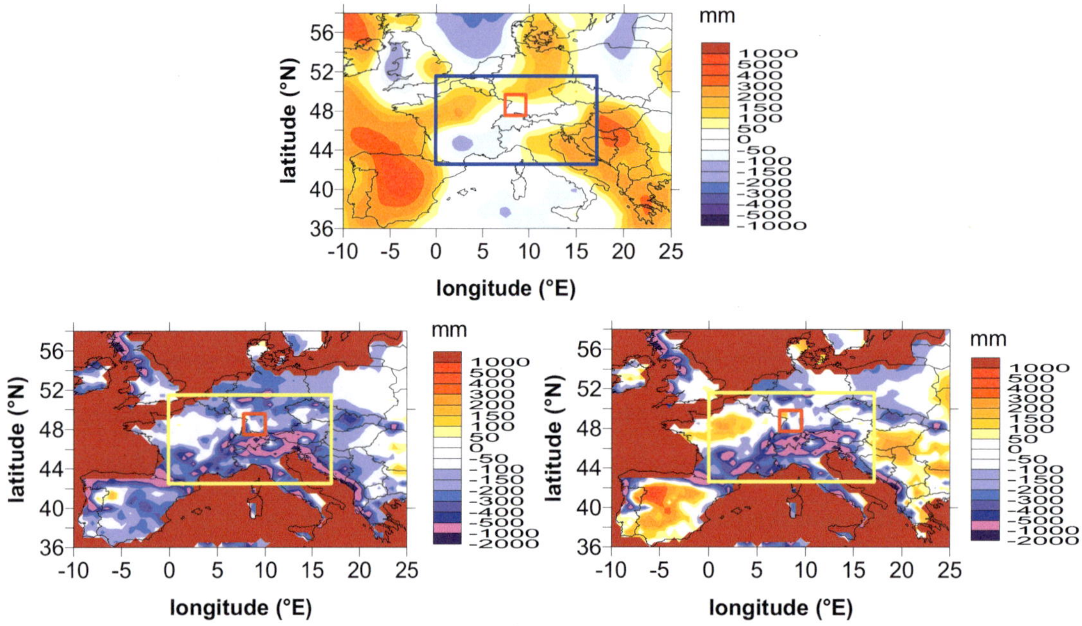

Figure 4.3: Difference in annual precipitation amount averaged over the period 1990 – 1999: difference between NCEP and ERA reanalysis (top), ERA and observations (bottom, left) and NCEP and observations (bottom, right). The red rectangle indicates the area of investigation and the blue/yellow rectangle the area of the alps domain.

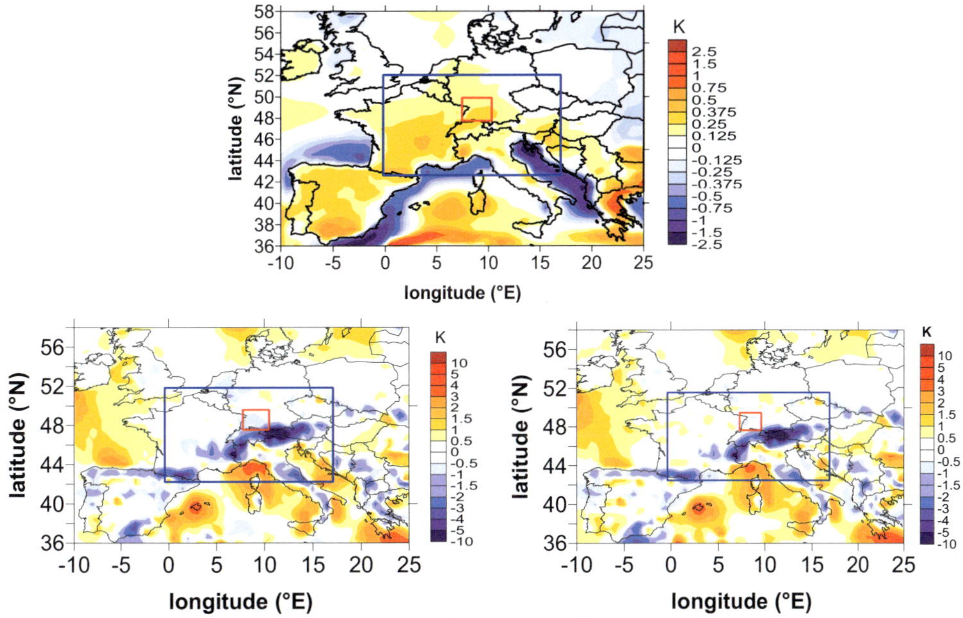

Figure 4.4: Difference in annual mean near-surface temperature averaged over the period 1990 - 1999: difference between 50 km simulations with NCEP and ERA (top), ERA and observations (bottom, left) and NCEP and observations (bottom, right). The red rectangle indicates the area of investigation and the blue rectangle the area of the alps domain.

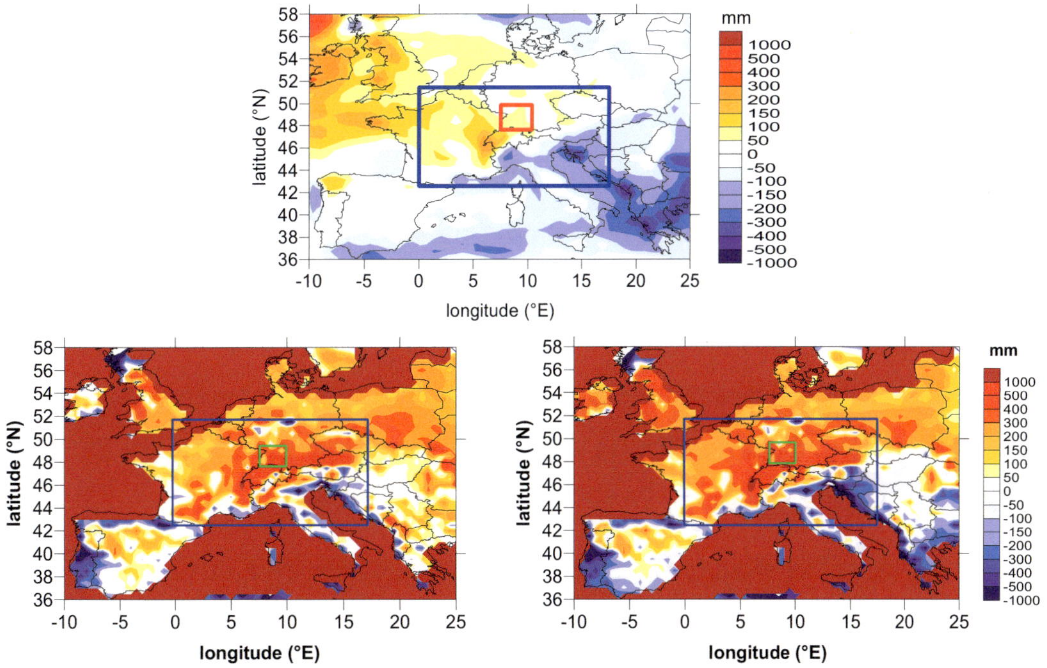

Figure 4.5: Difference in annual precipitation amount averaged over the period 1990 - 1999: Difference between 50 km simulations with NCEP and ERA (top), ERA and observations (bottom, left) and NCEP and observations (bottom, right). The green rectangle indicates the area of investigation and the blue rectangle the area of the alps domain.

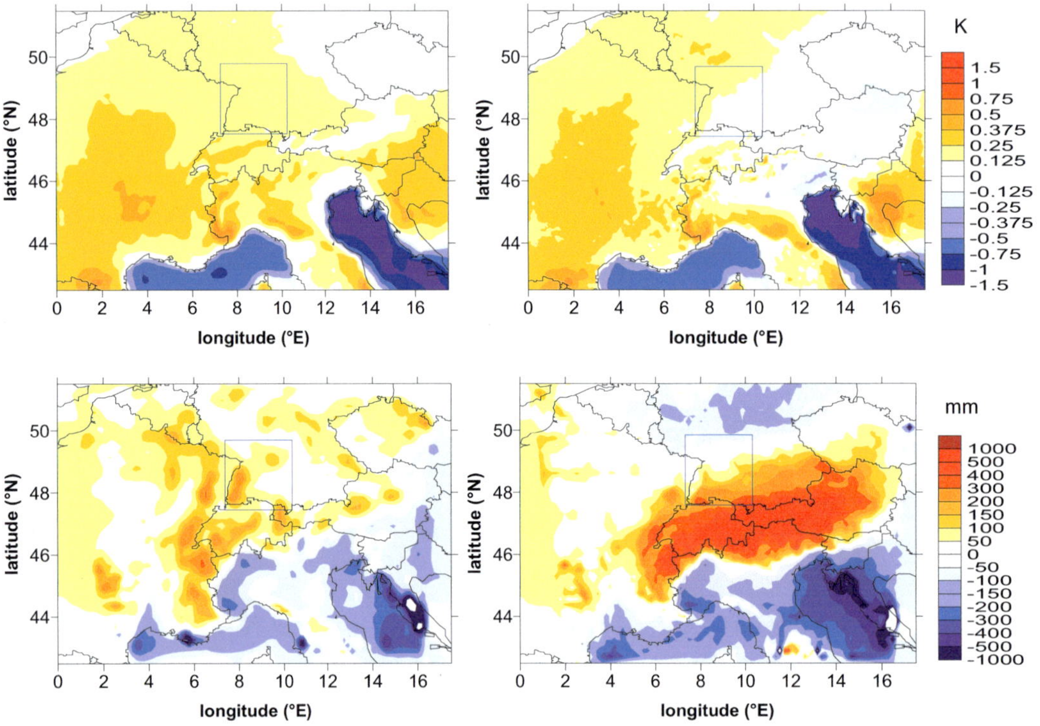

Figure 4.6: Difference between NCEP and ERA driven simulations of averaged annual mean near-surface temperatures (top) and averaged annual precipitation sum (bottom) for the period 1990 - 1999 for the 14 km run (left) and the 7 km run (right).

The temperature differences between the NCEP and the ERA simulations with 7 and 14 km resolution are smaller for higher resolutions (Fig. 4.6). The differences in precipitation amount increase with increasing resolution. As in the 50 km simulations, the NCEP simulation is warmer and yields more precipitation than the ERA simulation. The temperature differences between the simulations are even lower than in the 50 km runs (0.3 K). With increasing resolution, the differences in precipitation amount compared to the 50 km runs and the reanalysis data increase by up to 1000 mm over the Alps.

Differences observed between the two reanalysis data sets are passed on through the downscaling chain and are still visible in the high-resolution runs. The differences in temperature are smoothed out but the differences in precipitation prevail and are even strongly enhanced by higher resolutions.

4.2.2 Influence of initialisation

A third aspect which affects the model results besides driving data and model resolution is the initialisation of the model. The atmospheric initialisation is not as important as the soil initialisation because the memory of the soil lasts much longer. The soil initialisation can therefore exert an influence on the simulation results over many years. YANG et al. (1995) found that most land surface schemes need at least two years of spin-up time depending on vegetation type and initial soil moisture content. The initialisation of the soil temperature and the soil water content that is taken over from the reanalysis data is rather inadequate. The reanalysis data often contains less soil layers than the COSMO-CLM model and also the soil types that influence water and heat conductivity can differ between the driving model and regional model. Furthermore, a different soil model is used in COSMO-CLM and the models used to produce the reanalysis data. This implies that the profile of soil water content and soil temperature taken from the driving data can differ considerably from the profile which would be produced by COSMO-CLM itself for that grid point. In the literature several initialisation methods are recommended for soil temperature and soil water initialisation. One method is to use climatological values (RODELL et al., 2005). Such climatology for each day of the year was provided by the ETH Zürich for the 50 km grid (see section 4.1). To initialize the simulation, the climatological values for soil temperature and soil water content were extracted from the data set for the day of the year, which is equal to the initialisation day, and the ERA values were replaced by these climatological values. To determine the differences between the two initialisation methods, two simulations from 1988 to 2001 were performed: one with the original ERA initialisation and one with the climatological initialisation, each with a horizontal resolution of 50 km. The soil has nearly reached a steady state in both simulations after two years. At that point the root mean square deviation between the two simulations shows only small differences in soil moisture and soil temperature in the upper soil levels for all land points (<3 vol.-% and <0.6 K) (Fig. 4.7). An annual cycle with a higher root mean square deviation in summer than in winter can be observed. This is due to convection during the summer months. The small differences in soil temperature and soil moisture lead do a non-linear feedback with the development of convection and rainfall, which then leads to an enhancement of differences in soil moisture and soil temperature between the two simulations. This annual cycle persists over the whole simulation period and leads to random differences in the annual precipitation sum and in the annual mean near-surface temperature in the whole simulation area (Fig. 4.8). The root mean square deviation in annual mean near-surface temperature and evaporation between the two simulations diminishes during the first three years and stays constant afterwards. The root mean square deviation in precipitation stays constant over the complete period and shows no decrease during the first years (Fig. 4.9). It is not possible to determine, which of the initialisations gives better results because the differences in annual mean near-surface temperature are smaller than measurement accuracy after several years. Nevertheless, the use of the climatological values for initialisation is recommended for consistency reasons and is used for the simulations presented in the following sections.

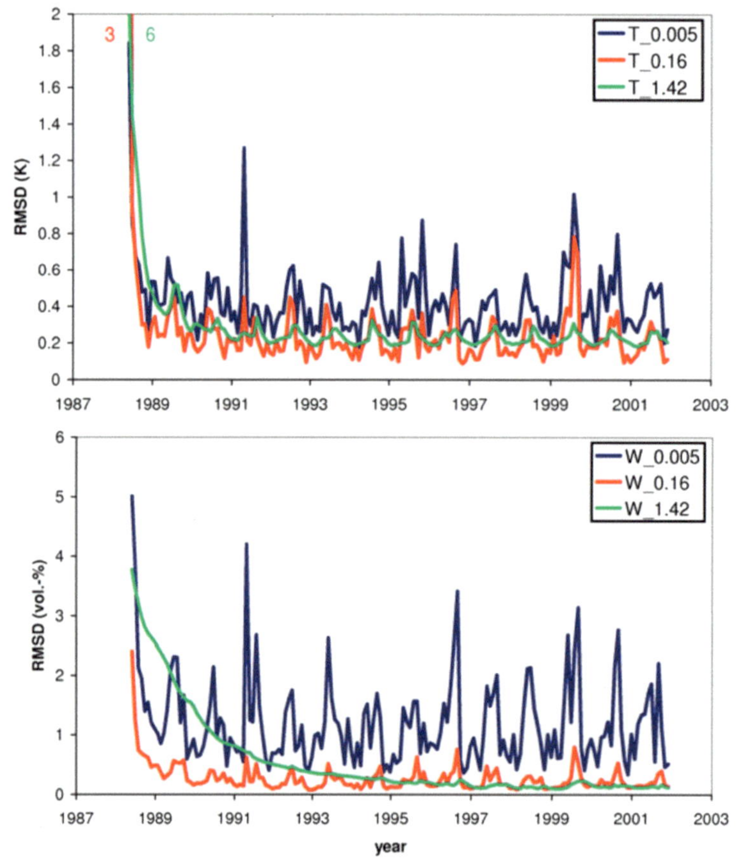

Figure 4.7: Root mean square deviation for all land points of soil temperature (top) and soil moisture (bottom) between simulation with original initialisation and climatological initialisation for different soil depth (blue: 0.005 m, red: 0.16 m, green: 1,42 m).

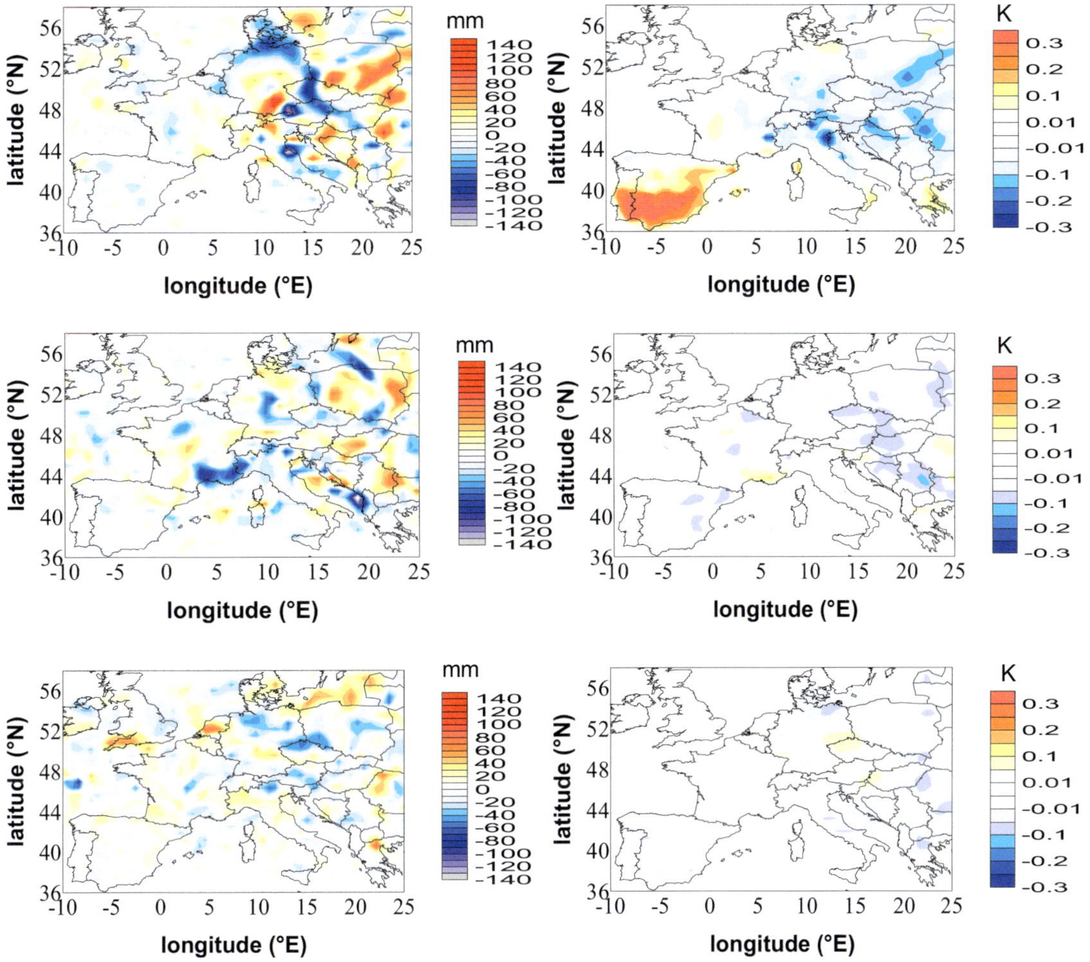

Figure 4.8: Differences between the simulation with ERA and climatological soil initialisation of annual precipitation amount (left) and annual mean surface temperature (right) for the year 1989 (top), 1995 (middle) and 2001 (bottom).

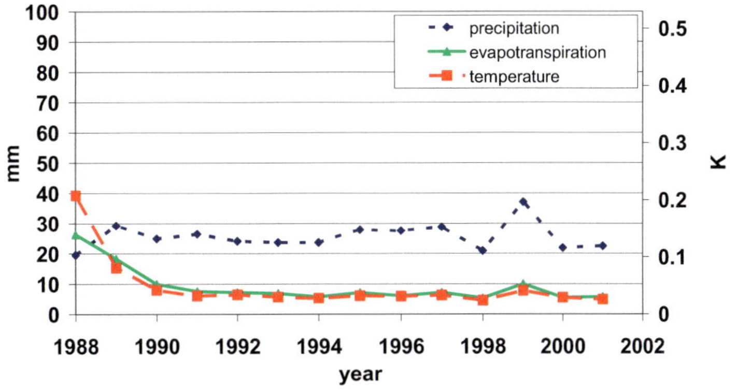

Figure 4.9: Root mean square deviation for all land points of total annual precipitation, annual evapotranspiration and annual mean near-surface temperature between simulation with ERA and climatological soil initialisation.

4.3 Sensitivity studies for Southwest Germany

4.3.1 Influence of simulation area

In the southern part, the area of investigation ends at the alpine foothills and it would be convenient to have the boundary of the simulation area in this region. It is well known that placing the model boundary in mountainous regions can be inconsistent with the boundary condition formulations and causes difficulties in the simulation of the vertical wind field because of the non-hydrostatic formulation of the model. Alternatively, the whole alpine region has to be simulated with corresponding CPU costs. To investigate the behaviour of the vertical wind field in the area of interest, two simulations were performed: one for an area without the Alps, referred to as no-alps-simulation, and one for a larger simulation area containing the Alps already referred to as alps-simulation (Fig. 4.1). Both simulations were driven by the 50 km simulations obtained with ERA-40 reanalysis as driving data and were run from June 1988 to December 2001. In comparing the two simulations cases with high precipitation amounts and either southerly or northerly winds are of interest. Therefore, March 1991 was chosen for a comparison of the two simulations. During this period very humid air is transported from the south over the Alps into the investigation area. The results for the 22nd of March 1991 are presented here as a case study. The findings are valid also for all other days during this period. On this day the vertical wind velocity in the alpine foothills is much higher in the no-alps-simulation than in the alps-simulation (Fig. 4.10). This vertical wind leads to condensation of humidity and increasing cloudiness. The condensed water is transported into the investigation area by the southerly winds, where it leads to an enhancement of precipitation compared to the alps-simulation. A comparison with gridded observations shows that this enhancement is unrealistic (Fig. 4.10).

In view of these unrealistic wind fields and precipitation patterns, the complete alpine region should be included in the simulation area. To determine how large the domain should be, Panitz (personal communication, 2007) performed a set of simulations with varying size of the simulation area for March 1991 to study the sensitivity of modelled total precipitation. The size of the model domain was the only model parameter being varied.

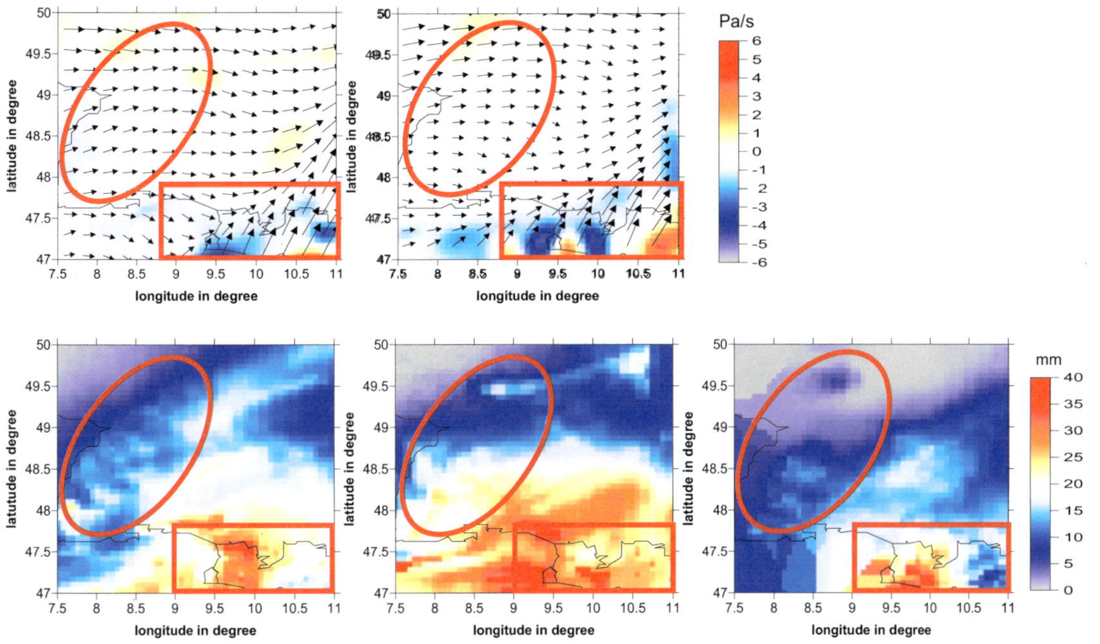

Figure 4.10: Wind field and vertical velocity in 700 hPa at 12 UTC on 22nd of March 1991 (top) and daily precipitation sum on that day (bottom) for the alps (left) and the no-alps simulation (middle) and for observation (right). Areas of strong precipitation reduction due to area enlargement are indicated in red.

Five of the different areas are shown in Fig. 4.11 and the results of the sensitivity studies for these areas are depicted in Fig. 4.12. The smallest region is the no-alps area. It intersects the most northern parts of the Alps to the south. Most striking is the unreasonable result of 1032 mm for the maximum value of monthly precipitation sum in the no-alps area, whereas the maximum gridded observation value is about 130 mm for March 1991. The observed area average is about 49 mm compared to 85 mm in the model. The size of model domain 4 exceeds that of the no-alps area by about 100 km in each direction. The area now includes the northern parts of the Alps. In comparison to the results for the no-alps area, the maximum monthly sum of total precipitation decreases considerably, but the value of 324 mm is still too high. The area average decreases slightly to 82 mm. The size of model domain 3 has been increased once more by 100 km in each direction and it encloses most parts of the Alps. Area 2 has essentially been elongated to the south to encompass the whole Alps, except the most eastern parts and Tuscany in Italy. The size of the largest area 1 exceeds that of the no-alps area by a factor of about eight and it now includes the whole Alps and also the Massif Central in France. The results for the total precipitation within the no-alps area derived from the simulations for domains 1 to 3 now seem to stabilize. The area averages are nearly identical, the maximum values vary within ± 7%, and the minimum values vary within ±16 %. However, COSMO-LM overestimates the observed monthly precipitation values for March 1991 - a tendency that is found especially for the months in winter and early spring (FELDMANN et al., 2007).

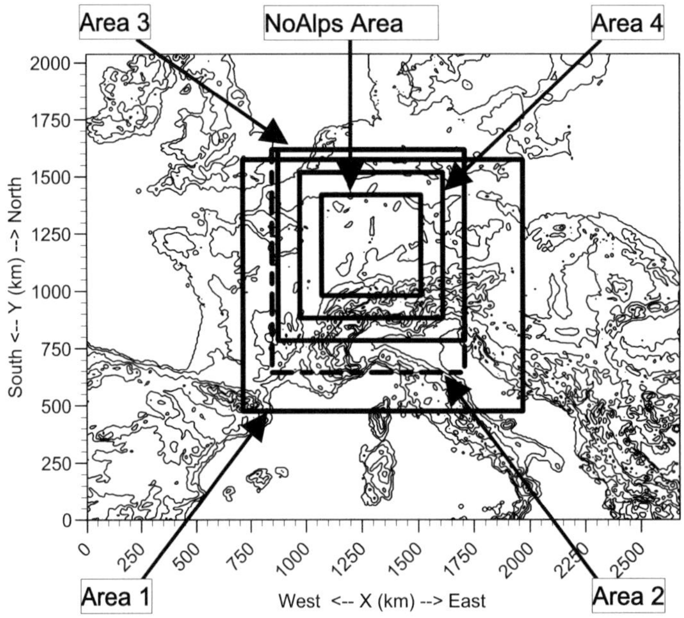

Figure 4.11: Model domains for sensitivity analyses of total precipitation (Panitz, 2007).

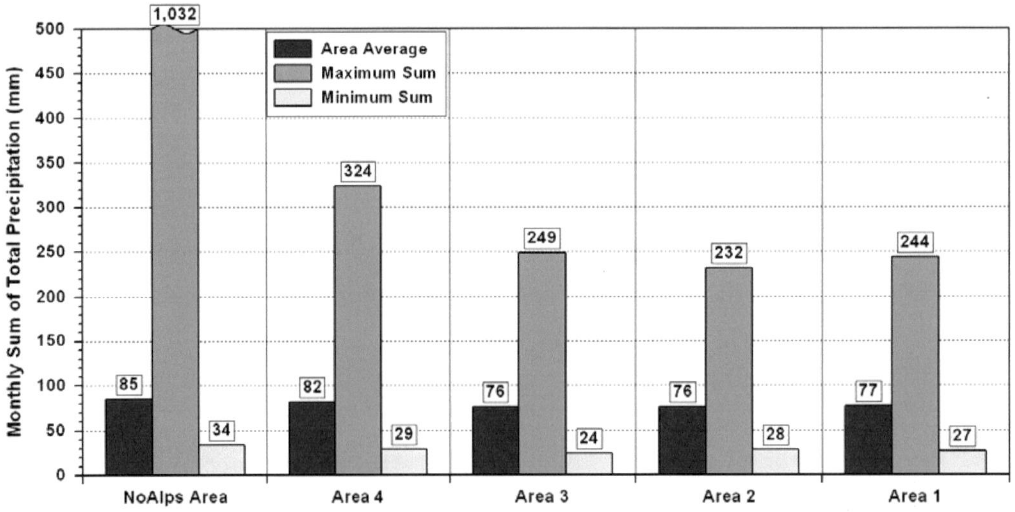

Figure 4.12: Area average, maximum and minimum sum of total precipitation in the no-alps area based on COSMO-CLM simulations with different sizes of model area for March 1991 (Panitz, 2007).

This sensitivity study makes it clear that the size of the model domain has to be considerably larger than the size of the actual investigation area - at least 200 km in each direction. Furthermore, it is necessary to include mountain ranges like the Alps if they are located rather close to the investigation area. For all simulations in the following chapters therefore area 1 was used, although area 2 or 3 would probably also have been suitable.

4.3.2 INFLUENCE OF HORIZONTAL RESOLUTION AND DRIVING DATA

Up to now, CPU time has been the limiting factor in high-resolution climate modelling. Therefore, it is worth studying whether higher horizontal resolution really results in better climate scenarios. In the standard COSMO model, four horizontal resolutions below 20 kilometres are available (km): 14, 7, 2.8, and 1. For 2.8 km and 1 km, the physics and dynamics change considerably compared to the 7 km and 14 km resolution (e.g. convection scheme and time integration scheme) and the CPU time needed is too high for standard use in high-resolution climate modelling. Therefore, this study focused on the comparison of simulations with 7 km and 14 km horizontal resolution.

Four simulations with 7/14 km resolution driven by the results of the 50 km runs with ERA-40 and NCEP data (already described in section 4.2) were used. This small ensemble of four simulations was compared to the observations described in section 4.1. Because the focus lies on precipitation, water cycle, and temperature, the grid scale, convective, and total precipitation was compared as well as the 2m-temperature, the water storage in the soil, the evaporation, the runoff, and the snow store to find the simulations with the most realistic results. The 2m-temperature and the total precipitation are the only variables, which can be evaluated qualitatively by observations. For all other variables only a quantitative comparison is possible due to the lack of adequate observations. For total precipitation and 2m-temperatures, the root mean square error (rmse) and the bias between the simulations and the station measurements were calculated for the annual precipitation sum and for the annual mean temperature. For precipitation, the rmse is lowest in the simulation with ERA driving data and 7 km resolution and highest in the simulation with NCEP driving data and 14 km resolution (Fig. 4.13). The rmse increases from higher resolution to lower resolution and from ERA to NCEP simulation. For temperature, an increase in rmse from NCEP to ERA driving data and an increase from 7 km to 14 km resolution (Fig. 4.13) can be observed. The smallest error for temperature is obtained with the NCEP/7 km simulation. The same behaviour can be found for the monthly values of precipitation and temperature (Fig. 4.14). The rmse is lowest for precipitation during the summer months and it can be observed that all four ensemble members are quite close to each other during this time. In winter, the spread of rmse between the simulations is larger and the absolute values of rmse are much higher. In winter, the precipitation seems to be influenced mainly by the driving data (with the differences already described) while in summer the model seems to produce its proper climate, where all four simulations give similar results. A similar behaviour was observed by FELDMANN et al. (2007). The rmse of temperature is highest in January for all simulations and decreases over the year until October, where the lowest spread between the four simulations is also found. A possible reason for this could be the proper climate produced by the model, above mentioned, which also influences the soil temperature. Because soil temperature is a slowly changing variable the agreement of the four simulations takes two months longer than for precipitation.

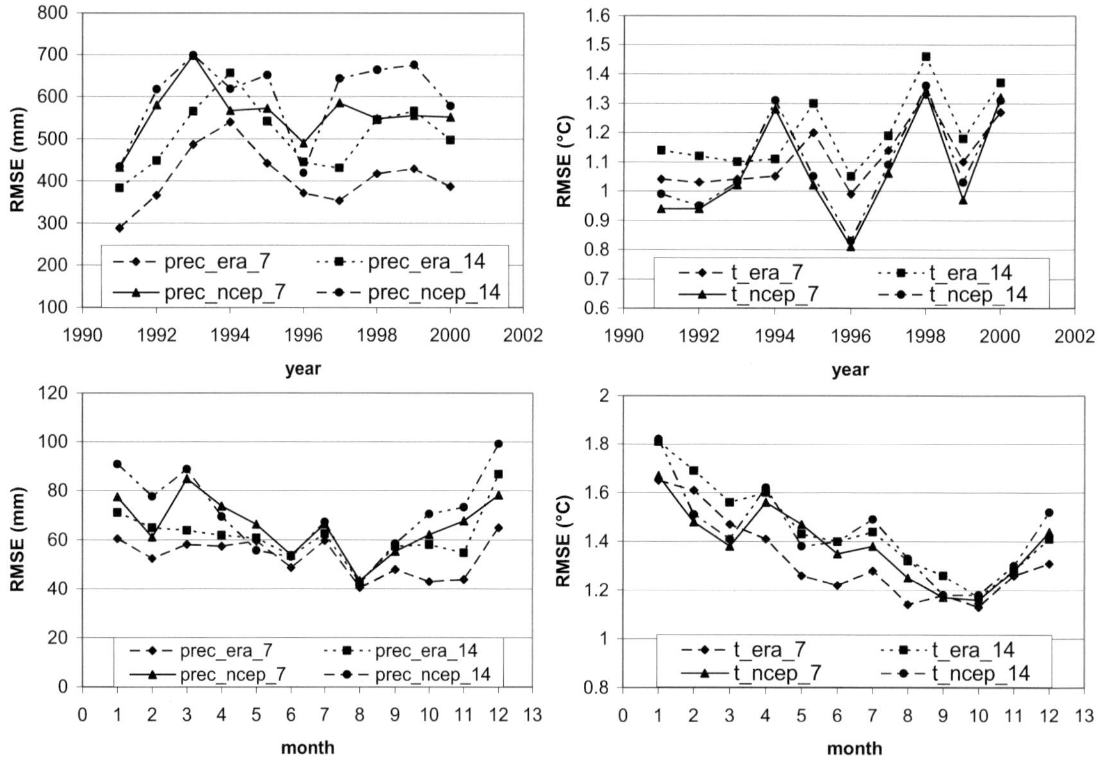

Figure 4.13: Root mean square error for precipitation (left) and 2m-temperatures (right) for the annual sum/mean (top) and for the monthly sum/mean (bottom) for the period 1991-2000.

To quantify whether the simulations over or underestimate the observations, also the bias for the annual and monthly values is presented (Fig. 4.14). All four simulations overestimate the yearly precipitation amount. Comparing the annual bias, the 14 km runs overestimate precipitation more than the 7 km runs and NCEP driving data more than ERA. The bias in annual mean temperature is lowest for the 7 km NCEP simulation. There is a negative bias over the whole decade and therefore an underestimation of temperature in all simulations. The bias of the monthly precipitation sums also shows an overestimation of precipitation by the simulations, which are strongest in winter and nearly zero in summer. As for the rmse, the spread of the precipitation bias is lowest in summer and highest in winter. The monthly temperature bias is highest in spring and decreases over the rest of the year. Over the whole year, there is a negative temperature bias in the monthly means of all four simulations.

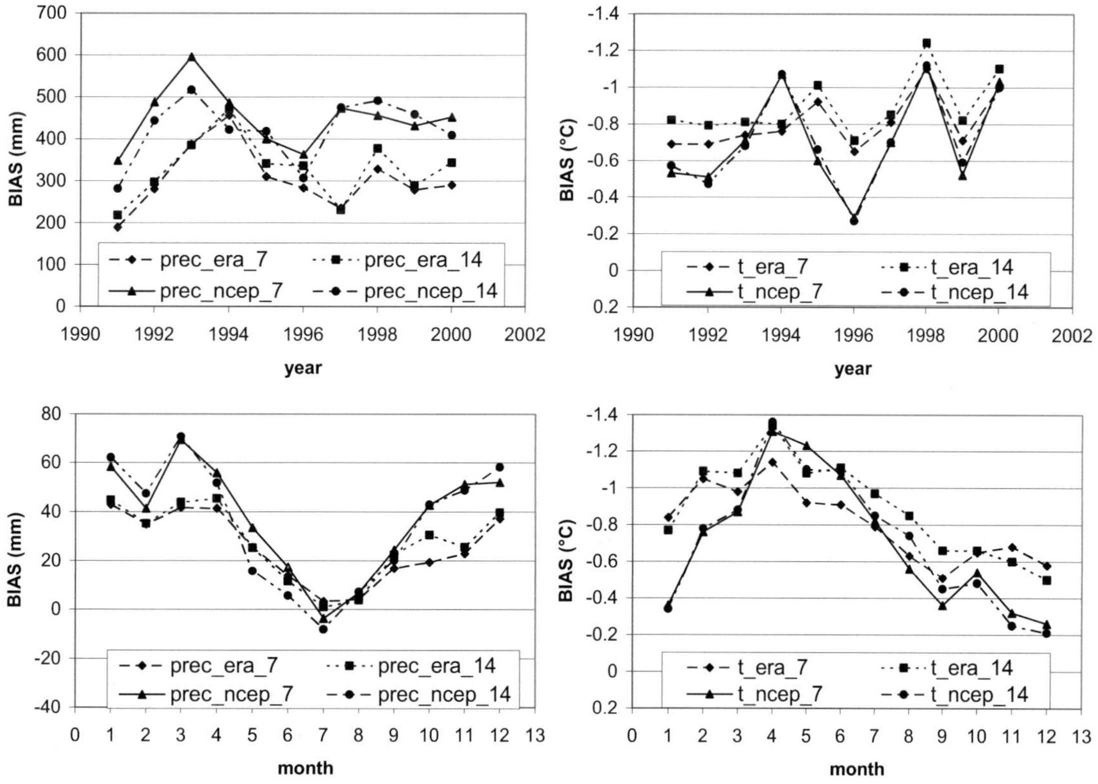

Figure 4.14: Bias between the simulations and the measurements for precipitation (left) and 2m-temperatures (right) for the annual sum/mean (top) and for the monthly sum/mean (bottom) for the period 1991-2000.

The simulations with the ERA driving data cause drier and colder conditions than those driven by NCEP data. As shown in section 4.2, this behaviour is already evident for the ten year mean in the 50 km runs that are used as driving data here and in the original reanalysis data. The best result for precipitation simulation compared to the measurements is obtained with the ERA/7km simulation. The results are not as clear for temperature as for precipitation. With respect to the annual means, best results are obtained for the NCEP simulation for both resolutions; for the monthly means, the best results are obtained for the ERA/7km simulation during the summer months. During the winter months, the NCEP simulations give best results. From this analysis it can be concluded that a resolution of 7 km should be used for simulations of this area and that ERA data is more suitable as driving data if the focus is on precipitation evaluation. For temperature, there is no large difference between ERA and NCEP data. The 7 km simulation with ERA driving data shows the smallest relative error.

The larger positive bias of precipitation in the 14 km simulation can be explained by a shift of the precipitation maximum from the ridge of the mountains towards the windward side. With regard to the area average of total precipitation one would have expected a lower bias in the 14 km runs because the area average is about 6-65 mm lower than the average for the 7 km runs which is closer to observations (Tab. 4.2). But some of the stations are situated to the west of the Black Forest, which is mostly of the time the windward side and this shift of maximum precipitation leads to an overestimation of annual precipitation sum at these stations (e.g. FRHBF, LAHR and BADENW) for the 14 km runs (Fig. 4.15) and in total to a much higher rmse than in the 7 km runs. For most of the stations to the east of the Black Forest (lee) or at its ridge, lower precipitation amounts are observed in the 14 km simulations than in the 7 km simulations, which underestimate the measurements (e.g. FELDBE, FST and TRBG).

	ERA/7km	NCEP/7km	ERA/14km	NCEP/14km
Total precipitation	1166	1260	1160	1224
Convective precipitation	351	358	313	332
Grid scale precipitation	815	902	847	892

Table 4.2: Area average of annual total, convective and grid scale precipitation sum for the investigation area averaged over the period 1991 – 2000.

The correlation coefficient of the linear fit between measurements and precipitation simulation shows much higher values for the 7 km runs (~0.5) than for the 14 km runs (~0.18) (Tab. 4.3). The intercept also shows lower values for the 7 km runs and the slope is closest to one for the 7 km simulations.

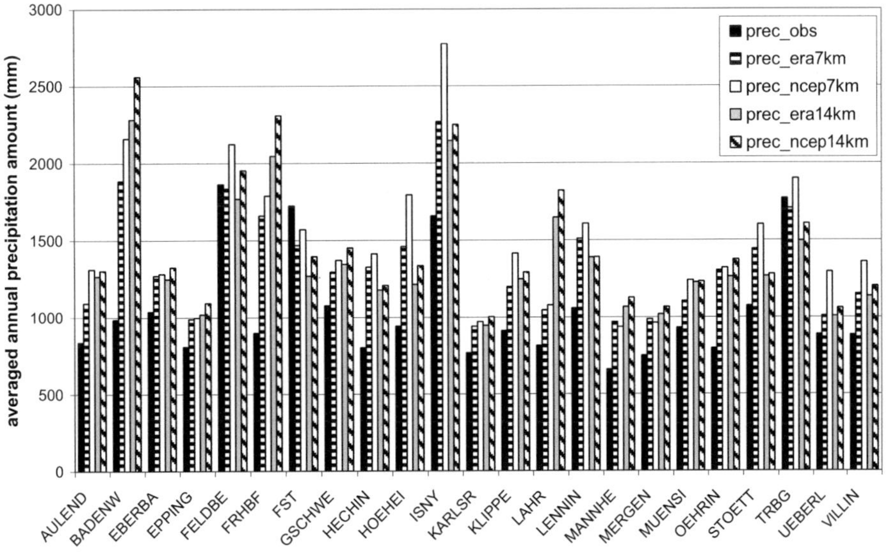

Figure 4.15: Annual precipitation amount averaged over the period 1991-2000 for the 23 observation sites (prec_obs) and the four COSMO-CLM simulations with different horizontal resolution (7 and 14 km) and different driving data (ERA, NCEP).

	a	b	r
ERA/7km	0.70	607	0.52
NCEP/7km	0.89	555	0.50
ERA/14km	0.46	889	0.19
NCEP/14km	0.51	926	0.17

Table 4.3: Linear correlation coefficient (r), slope (a) and intercept (b) of the least squares fit between observed and simulated annual precipitation amount averaged over the period 1991-2000 for the 23 observation sites and the four COSMO-CLM simulations with different horizontal resolution (7 and 14 km) and different driving data (era, ncep).

For orographically structured regions, it seems to be of advantage to use 7 km resolution in order to get the position of precipitation maxima correct. The amount of convective precipitation increases with increasing horizontal resolution by about 25-40 mm. The grid scale precipitation shows a decrease from 14 km to 7 km in the ERA runs of about 30 mm and an increase from 14 km to 7 km for the NCEP runs of about 10 mm (Tab. 4.2).

The influence of driving data and horizontal resolution on the water balance at the surface is shown in Tab. 4.4. The water balance is calculated as summed evaporation, runoff, and precipitation from January 1991 to January 2001 and averaged the balance over the whole investigation area. Evaporation shows quite similar results for all simulations and the evaporation sum decreases from the 7 km to the 14 km simulations. The surface runoff in the 7 km runs is higher than in the 14 km runs. This is possibly due to the higher precipitation maxima in the 7 km runs, which cannot be taken up by the soil and were added to surface runoff. Due to the higher precipitation amount in the NCEP runs, the runoff is higher in these simulations as well. Runoff from the soil increases from the 7 km to the 14 km resolution and from ERA to NCEP. The precipitation is not as intense in the 14 km runs as in the 7 km runs and can therefore infiltrate into the ground and lead to additional soil runoff compared to the 7 km runs. The total sum of precipitation increases with increasing resolution and increases from ERA to NCEP. For snow content and soil water content, the difference between the values at the first of January 1991 and at the first of January 2001 was calculated. The snow store and the soil water content increase during the simulations, the latter by 30 kg m^{-2} in the hydrologically active layers. An evaluation of the water balance is not possible due to the lack of observations for comparison. The additional precipitation amount observed in the NCEP simulations does not remain in the system. There is no markedly increased evaporation or rising soil water content in the NCEP simulations. The difference in the runoff between ERA and NCEP simulation is in the same range as the surplus of precipitation amount and has the same sign. This means that all additional precipitation in the NCEP simulations goes into runoff and is no longer available in the system. Some components of the water cycle like precipitation and runoff are strongly related to driving data and horizontal resolution and others, like soil water content and evaporation are less sensitive to changes in the horizontal resolution and the driving data.

Model run	Evapo-ration (sum)	Surface Runoff (sum)	Ground runoff (sum)	Total runoff (sum)	Precipi-tation (sum)	Difference of snow store	Difference of water content
ERA/7km	5675	1948	4126	6074	11665	4	29
NCEP/7km	5677	1988	5085	7072	12601	8	31
ERA/14km	5483	1720	4523	6244	11602	4	32
NCEP/14km	5588	1683	5165	6849	12240	9	27

Table 4.4: Water balance components at the surface in kg m^{-2} averaged over the investigation area summed up from January 1991 to January 2001.

4.3.3 Sensitivity with respect to varying parameterisations, adjustable parameters, and initialisation

Six 7 km simulations with varying parameterisations or adjustable parameters were performed with ERA-40 driving data in order to investigate the sensitivity of the model to those changes (Tab. 4.5):

1) Number of hydrologically active layers

In the standard COSMO model, the depth of hydrological active soil layers is restricted to 2.50 m (~ 7 layers). This means that below this depth no water can be transported upwards towards the soil layers near the surface and only gravitational transport downward is possible at the lower boundary. When the soil runs dry, there is no possibility to get water from deeper layers. Allowing water to rise from deeper soil layers could prevent this drying out. In nature, ground water rises and can be transported upwards to soil levels near the surface, which is not possible in the COSMO-CLM. Therefore, in one of the simulations all soil layers were considered as active soil layers, with only the behaviour of the lowest soil layer remaining unchanged.

2) + 3) Initial soil water content

It is generally assumed that the soil water content "forgets" its initialisation after some years. However, it is conceivable that the soil moisture content can run into different end states (Rodriguez-Iturbe et al., 1991), depending on the initial state. To see if this could also happen in these simulations and to investigate how long it takes the model to reach a steady state, two simulations were run: in one simulation 50% was added to the initial soil water content and in the other simulation 50% was subtracted from the initial water content.

4) Amount of greenhouse gases

In the COSMO-CLM, the CO_2 concentration is described by an increase following the A1B scenario between 1950 and 2100. For the past and present climate, this scenario describes the observed increase in CO_2 content. In the standard version of the COSMO model a constant CO_2 concentration of 330 ppm is used. To assess the difference over the decade, one simulation with this constant gas concentration was performed.

5) Convection scheme

The convection scheme was changed from the standard Tiedtke mass flux scheme to the Kain-Fritsch convection scheme (KAIN and FRITSCH, 1990) to determine the influence on precipitation and temperature.

6) Time integration scheme

The dynamics was changed by using the Runge-Kutta scheme instead of the Leapfrog scheme. The Runge-Kutta scheme is operationally used in the COSMO-DE, which runs on the 2.8 km scale (FÖRSTNER and DOMS, 2004). This scheme allows the use of more accurate advection schemes (fifth order horizontally).

The simulations where the time integration scheme or the convection scheme was changed show the largest differences compared to the reference run for annual mean temperature and annual precipitation amount (Fig. 4.16). The spread between the ensemble members is less than 0.3 degrees in annual mean temperature and less than 200 mm in annual precipitation sum. The observations do not lie within the ensemble spread for either temperature or precipitation. For all years, the ensemble overestimates the amount of precipitation by a minimum of 200 mm. For temperature, an underestimation by all ensemble members of minimum 0.6 degrees is found. This underestimation of temperature and overestimation of precipitation has also been observed in a study by BÖHM et al. (2006) for this region. The results for the change of water content and the change of active soil layers show only small variations compared to the reference run but they persist over the whole simulation period.

The variation in annual mean temperature is less than 0.01 degrees and the variation in annual precipitation sum is lower than 10 mm.

The simulation with constant CO_2 content shows slightly lower annual mean temperatures (up to 0.04 degrees without detectable trend over the simulation period) than the reference run and the change in annual precipitation sum is lower than 10 mm.

The simulation with the Kain-Fritsch convection scheme shows higher amounts for annual precipitation sums than the reference run (100 - 150 mm). The higher precipitation causes a decrease in temperature and therefore the annual mean temperature is up to 0.3 degrees lower than in the reference run.

Parameter	**COSMO-CLM (ref)**	**Ensemble (abbr.)**
Water content	Water content from driving model	+ 50% in all layers (min)
Water content		- 50% in all layers (max)
Active soil layers	Down to 2.50 m	All levels except the lowest (soil)
Greenhouse gas concentration	A1B scenario	Constant 330 PPM (co)
Convection scheme	Tiedtke scheme	Kain-Fritsch scheme (kf)
Dynamics	Leapfrog scheme	Runge-Kutta scheme (rk)

Table 4.5: Parameter setting for reference COSMO-CLM simulation and for ensemble simulations. The abbreviation for the ensemble members are given in the last column.

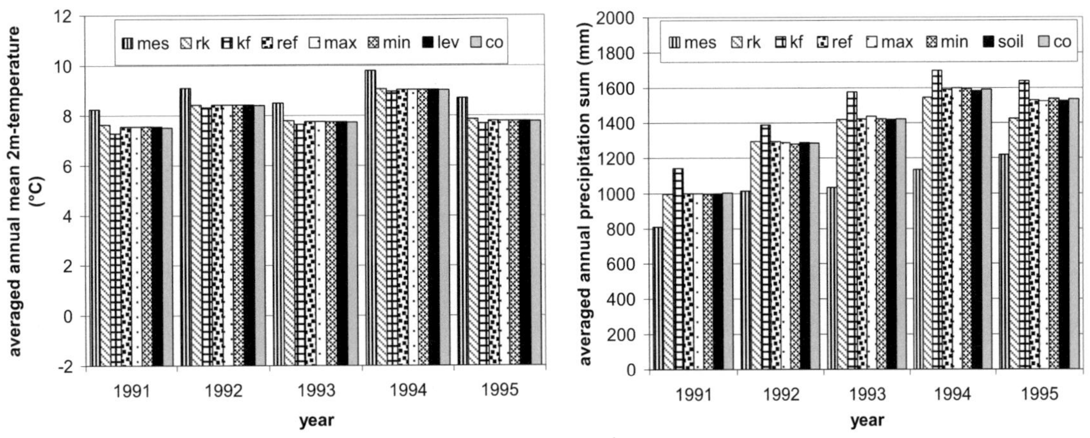

Figure 4.16: Annual mean of near-surface temperature (left) and annual sum of precipitation (right) for all ensemble members and observations averaged over the 23 stations.

In the simulation with the Runge-Kutta scheme, the precipitation amount decreases in the last two simulation years up to 100 mm compared to the reference run and the temperature variation is lower than 0.1 degrees.

To determine the spread between the monthly values they were averaged from 1991 to 1995 (Tab. 4.6). The monthly averages show that the ensemble spread is largest in June and July for 2m-temperature and from November to April for precipitation.

The change in soil moisture content has no marked influence on the annual course of temperature. Maximum differences are about 0.01 K. Differences of up to 4 mm in the monthly precipitation sum compared to the reference run occur during the vegetation period. This could be explained by the influence of soil moisture on the convective precipitation which also influences the temperatures.

The influence of the number of active soil levels on precipitation is observable during the whole vegetation period and reaches up to 5 mm per month compared to the reference run. Due to the water transport towards the root zone, slightly higher transpiration occurs and affects evapotranspiration and precipitation. The influence increases during springtime and is strongest in June and July. Afterwards, it decreases during autumn. The monthly variation of temperature is lower than 0.01 K. The influence on simulations due to soil moisture variation and soil level variation is therefore in the same range.

The simulation with the Runge-Kutta scheme influences temperature strongest in the time from October to March. In this period, the influence on precipitation compared to the reference run is strongest. This could be explained by the influence of the changed advection scheme, which has its strongest impact during the winter time. The differences between observations and simulation decrease compared to the reference run.

The underestimation of temperature by the simulation with constant greenhouse gas concentration is most significant during the time from October to March and is in the range of 0.03 degrees. The variation of precipitation in this simulation compared to the reference run is largest in summer and reaches up to 2.5 mm.

The differences in temperature between the simulation with the Kain-Fritsch convection scheme and the reference run are largest from May to August (up to 0.3 K). The simulation underestimates the temperature compared to the measurements to a higher extent than the reference run does.

Month	jan	feb	mar	apr	may	jun	jul	aug	sep	oct	nov	dec	year
	Temperature												
ref	-1.18	-0.84	-1.02	-1.13	-0.84	-1.14	-0.63	-0.48	-0.45	-0.35	-0.38	-0.72	-0.76
max	-1.18	-0.84	-1.02	-1.13	-0.84	-1.13	-0.63	-0.49	-0.45	-0.35	-0.38	-0.72	-0.76
min	-1.18	-0.84	-1.02	-1.13	-0.84	-1.13	-0.62	-0.48	-0.44	-0.36	-0.38	-0.72	-0.76
soil	-1.18	-0.85	-1.02	-1.13	-0.85	-1.13	-0.64	-0.49	-0.46	-0.36	-0.38	-0.72	-0.77
rk	-1.10	-0.75	-0.95	-1.14	-0.89	-1.20	-0.61	-0.49	-0.42	-0.26	-0.27	-0.61	-0.72
kf	**-1.27**	**-0.96**	**-1.11**	**-1.24**	**-1.04**	**-1.41**	**-0.94**	**-0.65**	**-0.57**	-0.35	-0.39	**-0.77**	**-0.89**
co	-1.21	-0.88	-1.04	-1.14	-0.85	-1.13	-0.65	-0.48	-0.46	**-0.38**	**-0.41**	-0.76	-0.78
Spread	0.17	0.21	0.17	0.11	0.19	0.28	0.33	0.17	0.15	0.12	0.16	0.16	0.13
	Precipitation												
ref	55.9	36.3	37.9	44.9	25.7	15.5	8.7	-0.3	17.3	13.7	28.1	40.7	324.3
max	55.8	36.4	37.6	43.7	25.3	12.9	**15.4**	0.9	16.1	13.8	28.0	40.7	326.6
min	56.0	36.2	37.5	44.5	25.4	12.9	11.1	1.7	16.7	13.4	27.7	40.6	323.6
soil	55.9	36.3	37.5	44.3	26.5	10.3	11.2	-1.2	16.5	13.9	27.7	40.7	319.5
rk	40.0	28.1	28.4	42.4	26.4	23.6	12.0	10.4	27.2	16.9	15.8	22.2	293.4
kf	**65.2**	**46.1**	**58.2**	**61.8**	**39.8**	**25.3**	0.1	1.5	**32.7**	**26.7**	**38.8**	**50.0**	**446.1**
co	55.9	36.3	37.1	45.2	24.8	14.2	11.3	1.9	17.0	13.4	27.6	40.5	325.0
Spread	15.2	18.0	29.8	19.4	15.0	15.0	15.3	**11.6**	15.8	12.8	23.0	27.8	152.7
Mean	54.9	36.5	39.2	46.7	27.7	16.4	10.0	2.1	20.5	16.0	27.7	39.3	336.9

Table 4.6: Difference of the monthly means and sums of 2m-temperature and precipitation between simulations and observations averaged over the years 1991 – 1995 for all ensemble members. Maximum values are in bold letters.

In contrast, the largest overestimation of precipitation is found in spring and autumn, where there is also a greater overestimation than in the reference run.

The ensemble member where the active soil layers were changed has the lowest impact (Fig. 4.17). There is only a bias of about 0.003 m^3 m^{-3} occuring in the originally active layers in the first winter period. In the seventh soil layer, formerly the lowermost active layer, a small annual variation of the bias of about 0.001 m^3 m^{-3} can be observed, with a maximum in the late autumn. This corresponds to the findings above: the change of active soil layers has an influence over the vegetation period, but the differences in soil moisture are negligible.

During the first six months of the simulations, the bias decreases for the two simulations with changed initial soil water content. After that time, the bias compared to the reference run is lower than 0.01 $m^3 m^{-3}$. The simulation with decreased soil water content shows a slow recovery over the first six months while the adaptation of water content in the simulation with increased soil water content happens essentially during the first month of the simulation and is mainly done by the production of runoff (Fig. 4.17). In the simulation with decreased soil water content, the lower layers fill up within one month after the upper layers were filled and the gravitational pressure becomes large enough. The evolution of soil water content is similar to the behaviour of the reference simulation for both ensemble members. There is no splitting up in different final states. When the soil water content has reached an equilibrium state after the first year of simulation, its influence on precipitation and temperature is rather small. Therefore, the choice of initial soil water content and number of active soil layers is not crucial for this investigation area and this initialisation date.

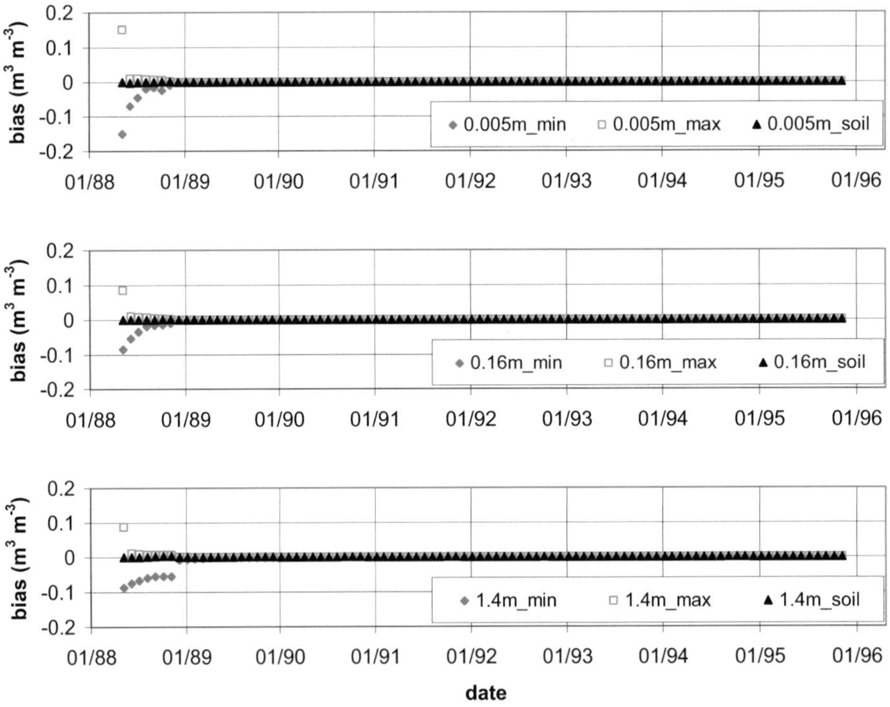

Figure 4.17: Bias of soil moisture content of simulations with changed soil moisture content (min, max) and different hydrological active layers (soil) compared to the reference

run averaged over the whole reference area for different soil layers for the first day of every month at 0 UTC. Labels indicate the depth of the main level of the layer.

To get an overview of the differences between the reference run and the simulations with the Kain-Fritsch and the Runge-Kutta schemes, the average of the annual means or sums of temperature and precipitation for the whole reference area over the period 1991 – 1995 was calculated and the results of the reference run were subtracted (Fig. 4.18). The results for the simulation with the Kain-Fritsch scheme show an increase in precipitation amount in all areas except on the western slopes of the mountains (which is the windward side most of the time) where a decrease is found (Fig. 4.18). This can also be seen in the temperature which decreases least to the west of the mountains. As discussed in subsection 4.3.2, precipitation is underestimated on the eastern slopes of the mountains and overestimated on the western slopes, which is slightly corrected by the use of the Kain-Fritsch convection scheme. However, the overestimation of precipitation with this scheme is higher than in the reference run.

For the Runge-Kutta scheme the largest differences from the reference run occur in mountainous regions (Fig. 4.18). The precipitation in the windward side of the mountains considerably decreases in this simulation and increases in the lee. In the alpine foothills annual precipitation sum is strongly reduced in this simulation. This behaviour agrees better with the gridded precipitation data than the reference run does. A warming of up to 0.3 K occurs in the south of the investigation area and a cooling of up to 0.15 K in the north. Since the computational costs do not increase, the Runge-Kutta scheme could be a possible alternative to the Leapfrog scheme for climate simulations.

In the simulation with constant greenhouse gas, the temperature decreases in most of the area compared to the reference run. The change in precipitation shows no evident structures and its sign changes more or less randomly throughout the area (Fig. 4.18).

In summary, changing the convection scheme and the time integration scheme has considerable impacts on the simulation of precipitation and temperature. The change in the greenhouse gas concentration shows the expected decrease in temperature and use of the increasing green house gas concentration given by the A1B scenario is recommended for the simulation of past decades.

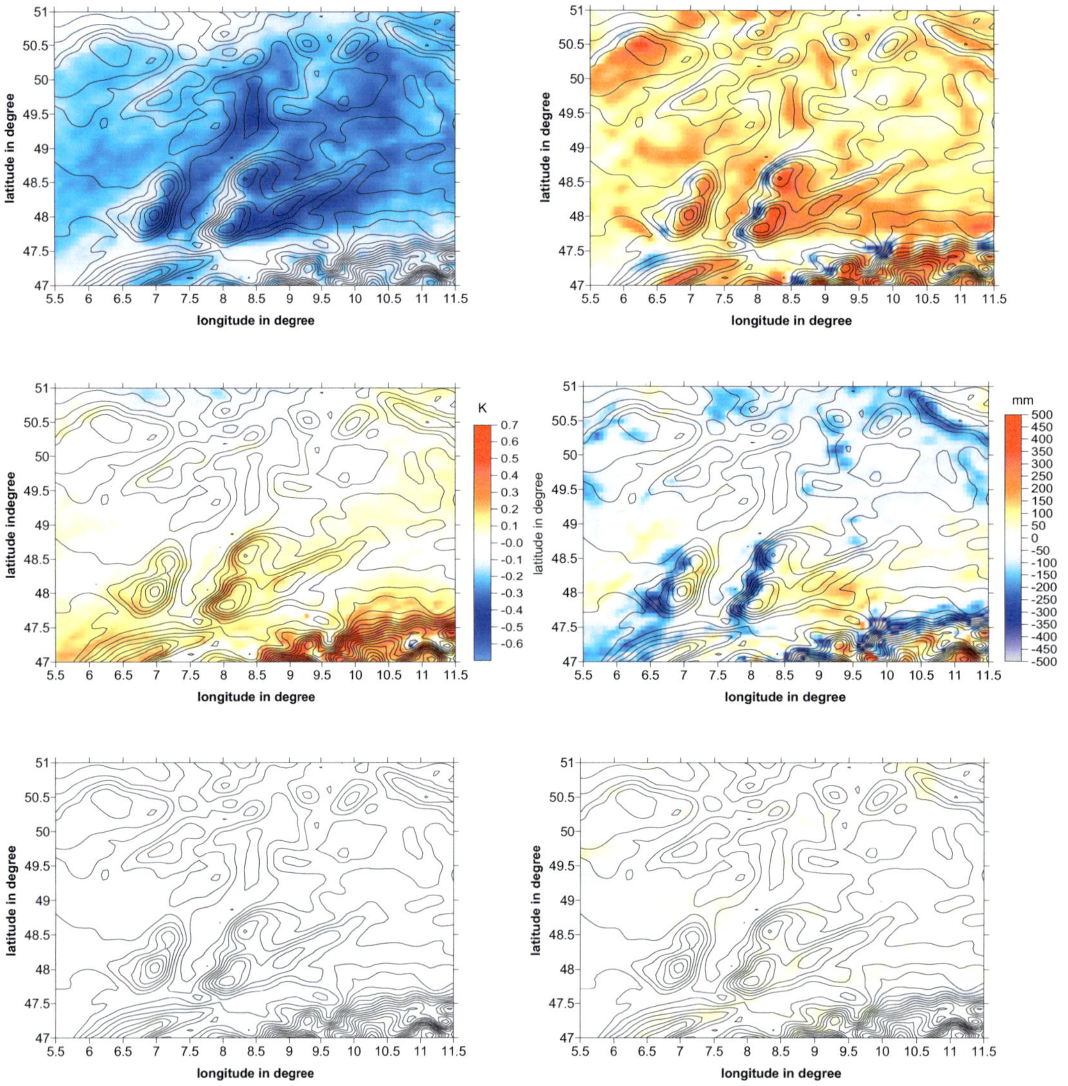

Figure 4.18: Difference between simulations with the Kain-Fritsch scheme (top), the Runge-Kutta scheme (middle) and constant greenhouse gas concentration (bottom) and the reference run for annual mean near-surface temperature (left) and annual precipitation sum (right) averaged over the period 1990 – 1995.

4.3.4 Comparison of observed and modelled trends of near-surface temperature between 1991 and 2000

The observed and modelled trends of the near-surface temperature for the decade 1991 to 2000 show a marked increase of the surface temperature at many sites in Southwest Germany (Fig. 4.19). Due to the shortness of this time interval, however, the significance of these trends is lower than for the trends between 1931 and 2000 (KLIWA, 2005). The modelled trends presented here were computed using the 7 km domain with ERA-40 driving data. The trends were calculated from annual average temperatures using the least squares method: their significance was calculated with the Mann-Kendall test (Rapp and Schönwiese, 1996). The orders of magnitude of the observed and the modelled trends agree reasonably well, both of them being mainly in the range between 0.1 and 0.9 K/decade. This is the same magnitude as those in the time span between 1931 and 2000 and shows that a considerable part of the warming took place during the last decade of the 20th century. Observations and model agree also with respect to the relatively strong warming that occurred in the Rhine valley (KLIWA, 2005), the western slopes of the Black Forest (Fig. 4.19), and in the Kraichgau region (Fig. 4.19), whereas modelled trends and/or significances are smaller than observed in the eastern parts (Fig. 4.19) of the Black Forest and extended portions of the Swabian Jura.

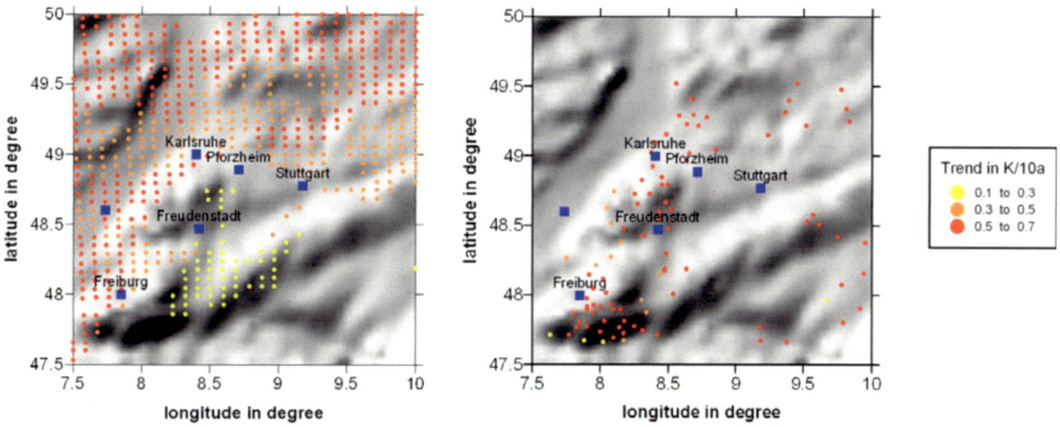

Figure 4.19: Modelled (left) and observed (right) trends of near-surface temperature between 1991 and 2000.

On the other hand, there are also some differences: some of the observed warming trends in the east of the state of Baden-Wuerttemberg are not reproduced by the model and modelled trends are - to a varying degree - smaller than the observed ones. Since smaller trends result in smaller significance, the significance of the modelled trend is smaller than the significance of the observations in this region.

4.4 Summary

For high-resolution climate simulations for Southwest Germany the simulation area should include the entire alpine region and the results for precipitation improve when using 7 km horizontal resolution instead of 14 km. The smallest bias to observed temperature and precipitation is found using ERA-40 data as driving data. In the time scales considered, changes in initial soil water content or in the number of active soil layers affect the simulation results only during the first year of simulation and can be neglected afterwards for the high-resolution simulations. For the 50 km simulations, initialisation from climatological soil moisture and soil temperature data is recommended. The model is quite sensitive to the change of the Tiedtke convection scheme to the Kain-Fritsch scheme, which leads to an overestimation of precipitation in comparison to observations. Model results are also quite sensitive to changing the time-integration scheme. The Runge-Kutta scheme seems to be an alternative to the Leapfrog scheme that has been standard in the COSMO-CLM up to now. It considerably reduces the precipitation amounts in the windward side of the mountains, which are generally overestimated by COSMO-CLM.

The water balance of the soil shows components, which are quite sensitive to the change of driving data and horizontal resolution (precipitation, runoff) and components, which are not sensitive to these changes (soil water content, evaporation).

For high-resolution simulations, the influence of varying lateral boundary conditions on simulation results is much larger than the change of horizontal resolution, initial conditions, physical parameterisations and time integration schemes. This is in accordance to other studies focusing on different horizontal resolutions and time scales (Anthes et al., 1989). The differences in temperature and precipitation already contained in the reanalysis data are passed on through the downscaling chain to the high-resolution simulations and are amplified for precipitation and weakened for temperature. The model is able to reproduce temperature trends with the suitable model setup determined in the section before by ensemble simulations.

5 Application and improvement of the land-surface scheme VEG3D

The soil-vegetation model VEG3D was initially developed by SCHÄDLER (1990). Later, a snow module (GRABE, 2002) and a root distribution function were included (LENZ, 1996). It is a multi-layer soil model with an explicit vegetation layer and it is extensively described in SCHÄDLER et al. (1990), BRAUN (2002) and GRABE (2002). A summary of the most important features is given in subsection 6.1.2.

Two major modifications of the model have been implemented and evaluated in this thesis. The first one is the implementation of a parameterisation of freezing and melting processes in the soil, which is indispensable for climate simulations and was missing in VEG3D up to now. The second modification is the consideration of different soil types within one soil column; up to now, only one soil type per soil column was allowed. The implementation and the results of stand-alone simulations with the new soil freezing parameterisation driven by observed data are described in section 5.1, and in section 5.2, stand-alone simulations with vertically varying soil types are presented.

5.1 Parameterisation of freezing and melting processes in the soil

For climate simulations, the lack of freezing and melting processes in the soil can lead to strong biases in soil temperature and soil water content. Furthermore, the simulated near-surface temperature can be too low compared to observations in winter (Viterbo, 1999) and more water penetrates into the soil if soil freezing is not taken into account (YU and SHIHUA, 2002). In addition, a strong influence on the development of the boundary layer by soil freezing processes can be observed (VITERBO, 1999). Therefore, in many of the land surface schemes, soil freezing processes were implemented over the past ten years (e.g. ISBA (BOONE et al., 2000), Sib2 (LI and KOIKE, 2001), BASE (SLATER et al., 1998)). Parameterisations of freezing and melting processes with different complexity have been developed; the Project for Intercomparison of Land-surface Parameterisation Schemes

(PILPS) has compared the ability of 21 land surface schemes to simulate soil freezing and snow coverage (LUO et al., 2002). It was found that the consideration of freezing and melting processes improves the simulation of soil temperature and its variability at seasonal and interannual scales. Due to the variety of freezing parameterisations already available, no new soil freezing scheme for VEG3D has been developed. Instead, different approaches from other land surface schemes have been combined into a suitable parameterisation.

Soil ice has influence on those parts of VEG3D where the soil temperature and the soil water content are calculated. The influence of soil ice on heat capacity and thermal conductivity for soil temperature calculation was adopted from the work of JOHANSEN (1975). The change in soil water transport was taken from the Common Land Model (DAI et al., 2001) and the change of soil temperature and soil water content due to energy release and absorption by freezing and melting was taken from the COSMO model (DOMS et al., 2005).

5.1.1 CALCULATION OF SOIL TEMPERATURE IN FROZEN SOIL

Soil ice is accounted for in the thermal part of VEG3D through its influence on the heat capacity and thermal conductivity in the heat conduction equation for the soil temperature T_B:

$$c_B(\theta)\frac{\partial T_B}{\partial t} = \frac{\partial}{\partial z}\left(\lambda_B(\theta)\frac{\partial T_B}{\partial z}\right). \tag{5.1}$$

The heat capacity of the soil c_B is the sum of the heat capacities of dry soil $c_{B,t}$, water c_w and ice c_e, each weighted by its fractional amount of soil volume:

$$c_B(\theta_w, \theta_e, \theta_s) = (1-\theta_s)c_{B,t} + \theta_w c_w + \theta_e c_e . \tag{5.2}$$

The heat capacity of air is neglected here because its value is much smaller than those of the three other heat capacities and it is assumed that the whole pore volume can be filled up by water. This means that, for the determination of c_B, only the volumetric soil water content θ_w, the volumetric ice content θ_e, and the volumetric saturation water content θ_s

need to be known. The thermal conductivity of the soil λ_B is described by the approach from Johansen (1975):

$$\lambda_B(\theta_w,\theta_e,\theta_s) = K_e(\theta_w,\theta_e,\theta_s)(\lambda_{B,s} - \lambda_{B,t}) + \lambda_{B,t}\,. \tag{5.3}$$

$\lambda_{B,t}$ is the thermal conductivity of dry soil and K_e is the Kersten number. The soil conductivity of saturated soil $\lambda_{B,s}$ is described for unfrozen soil as

$$\lambda_{B,s} = k_s^{\,(1-\theta_s)} \cdot k_w^{\,\theta_s}\,, \tag{5.4}$$

and for frozen soil as

$$\lambda_{B,s} = k_s^{\,(1-\theta_s)} \cdot k_e^{\,(\theta_s-\theta_w)} \cdot k_w^{\,\theta_w}\,. \tag{5.5}$$

The thermal conductivity of saturated soil depends on the effective solids thermal conductivity k_s, the thermal conductivity of ice k_e (2.2 W m^{-1} K) and the thermal conductivity of water k_w (0.5 W m^{-1} K). The thermal conductivity k_S depends on the quartz content q, the thermal conductivity of quartz k_q (7.7 W m^{-1} K) and the thermal conductivity of minerals in the soil k_0 (2.0 W m^{-1} K):

$$k_S = k_q^{\,q} k_0^{\,1-q}\,. \tag{5.6}$$

The Kersten number is determined as

$$K_e = \begin{cases} \left.\begin{array}{ll} \log\dfrac{\theta_w}{\theta_s} + 1 & \textit{fine soil } (p < 2\mu m) \\ 0{,}7\log\dfrac{\theta_w}{\theta_s} + 1 & \textit{coarse soil } (p > 2\mu m) \end{array}\right\} \textit{unfrozen soil} \\ \left.\dfrac{\theta_w + \theta_e}{\theta_s}\right\} \textit{frozen soil} \end{cases}\,. \tag{5.7}$$

For unfrozen soil, it distinguishes between fine and coarse soil, depending on particle size p.

5.1.2 Parameterisation of soil water transport in frozen soil

The soil water transport is described by the Richards equation

$$\frac{\partial \theta(z,t)}{\partial t} = \frac{\partial}{\partial z}\left[K(\theta,z)\left(\frac{\partial \Psi(\theta,z)}{\partial z} - 1\right)\right] - r_w(\theta,z). \quad (5.8)$$

r_w describes the water up-take by the roots. K is the hydraulic conductivity, Ψ is the matric potential and θ is the volumetric soil water content. For the description of hydraulic conductivity and matric potential, three parameterisations are available in VEG3D. Only the later used Van Genuchten parameterisation (1980) is discussed here:

$$\Psi(\theta) = \Psi_s\left[(S)^{-\frac{1}{m}} - 1\right]^{\frac{1}{n}}, \quad (5.9)$$

$$K(\theta) = K_s\sqrt{S}\left[1 - \left(1 - (S)^{\frac{1}{m}}\right)^m\right]^2, \quad (5.10)$$

$$S = \frac{\theta - \theta_r}{\theta_s - \theta_r}, m = 1 - \frac{1}{n}, n = \lambda + 1. \quad (5.11)$$

Depending on the soil type, the saturation matric potential Ψ_S, the hydraulic conductivity at saturation K_S, the residual water content θ_r, the saturation water content θ_S, and the dimensionless pore size distribution index λ are described by parameter data sets from RAWLS and BRAKENSIEK (1982). In the presence of soil ice, the volume for soil water transport is considerably reduced and the fractional parameter S is replaced by

$$S_e = \frac{\theta - \theta_r}{P_e}, \quad (5.12)$$

$$P_e = \theta_s - \theta_r - \theta_e. \quad (5.13)$$

The reduced volume available for soil water transport in the presence of soil ice is considered by the introduction of an effective porosity P_e, which is lowered by the volumetric ice content compared to the porosity without soil ice. The fractional parameter S_e then refers to the ice free soil volume.

For the three parameterisations BROOKS & COREY (1964), CAMPBELL (1974) and VAN GENUCHTEN (1980), the hydraulic conductivity and the matric potential remain unchanged. Only the new effective porosity is considered.

When the difference between the saturation water content and the soil ice content in a soil layer and the adjacent layer is lower than 0.13, the hydraulic conductivity is set to zero in this soil layer in accordance with BLOOMSBURG and WANG (1969): there is no water transport from or to the adjacent soil layers.

For the calculation of the infiltration rate, the sum of soil water and soil ice content is considered and the transpiration by the roots is restricted to the liquid water content.

5.1.3 ENERGY EXCHANGE DURING FREEZING AND MELTING PROCESSES IN THE SOIL

Before soil temperature can decrease under the freezing point T_0 (273.15 K), all water except the residual water content has to be frozen. In case of melting, the soil temperature can only rise above the freezing point after all soil ice within the soil layers is melted. The change of soil temperature and soil water content is calculated without explicit consideration of freezing and melting processes. A correction is done afterwards by diagnostically determining the energy available for freezing and melting processes using

$$\Delta E = c_B \left(T_B - T_0\right). \tag{5.14}$$

When $T_B < T_0$, the soil water is frozen; when $T_B > T_0$, available ice is melted. The maximum possible change in soil water content is therefore described by

$$\Delta\theta_{w,\max} = -\Delta\theta_{e,\max} = \frac{\Delta E}{L_f \rho_w}, \tag{5.15}$$

where L_f is the heat of fusion and ρ_w is the density of water. The observed change in water content depends on the available soil water content:

$$\Delta\theta_w = Min\left(\Delta\theta_{e,\max}, \theta_e\right) \; if \; \Delta\theta_{e,\max} < 0, \tag{5.16}$$

$$\Delta\theta_w = -Min\left(\Delta\theta_{w,\max}, \theta_w - \theta_{res}\right) \; if \; \Delta\theta_{e,\max} > 0. \tag{5.17}$$

The final soil temperature is calculated as

$$T_B = T_0 + (\Delta\theta_e - \Delta\theta_{e,\max})\frac{L_f \rho_w}{c_B}. \tag{5.18}$$

When all water except the residual water content has been frozen, the soil temperature can decrease under the freezing point.

5.1.4 Evaluation of freezing and melting parameterisation in the soil

The parameterisation of freezing and melting processes in the soil has been evaluated with data from the boundary layer field site at Falkenberg (NEISSER et al., 2002), which belongs to the Observatorium Lindenberg and is operated by the DWD. A detailed description of the measurement site can be found in section 6.2.

The water content is observed in six depths (in m): 0.08, 0.15, 0.3, 0.45, 0.6, 0.9. The soil temperature is measured in the following 12 depths (in m): 0.05, 0.1, 0.15, 0.2, 0.3, 0.45, 0.5, 0.6, 0.9, 1.0, 1.2, 1.5. There are only a few other observation sites that have such high-resolution vertical profiles of soil moisture and soil temperature available over several years. As atmospheric driving parameters, 2m-temperature and specific humidity, as well as downward long- and shortwave radiation, pressure and wind speed are available every ten minutes.

For the evaluation, the stand-alone version of VEG3D has been used with eight soil layers with the following depth of main levels (m): 0.01, 0.03, 0.07, 0.15, 0.31, 0.63, 1.42, 2.55. It is run as a single column model with land use type grassland and soil type loamy sand in the upper five soil layers and loam in the lowermost three soil layers. Soil temperature, soil water content and heat fluxes at the surface are simulated by the model and can be compared to observations. The observed soil water content and soil temperature was interpolated onto the depth of the main levels of the model. Therefore the soil water content can be evaluated for the levels from 0.07 to 0.63 m and the soil temperature for the levels from 0.07 to 1.42 m. A comparison with observations has been done for the winter season 2002/2003, with initialisation at the first of October, 2002, and for one year starting from first of June 2003.

For non-frozen periods, VEG3D simulates soil water content and soil temperature well and we observe correlation coefficients of 0.98 for temperature and 0.84 to 0.9 for soil water content. The largest differences between observed and modelled soil temperature and soil water content are in the range of 1.3 K and 5 volume percent and are found in deeper soil layers.

For the winter period 2002/2003, the episodes of soil freezing are displayed correctly by the model with the new parameterisation and soil temperatures accord more accurately with the observed ones than those without consideration of soil freezing (Fig. 5.1). Strong unrealistic oscillations, which are observed without soil freezing during periods with temperatures around freezing point, are eliminated with the new parameterisation; too strong cooling of the soil during some episodes is also prevented. The improvement in soil temperature simulation with the new scheme is also evident in deeper soil layers where no freezing occurs. The influence of soil freezing on temperature vanishes some days after the last freezing period and soil temperatures at the end of the simulation period for the simulation with and without soil freezing are quite similar.

The absolute values of soil water content can only be compared during the non frozen episodes due to the observation method with Time Domain Reflectometry (DALTON and VAN GENUCHTEN, 1984), which is not able to determine the soil water content of frozen soil correctly. But the freezing period can be determined by the decrease in observed soil water content. The periods of freezing are well represented by a decrease of soil water content in the layers, with main levels in 7 and 31 cm (Fig. 5.2). At the depth of 15 cm, one period of melting with increasing soil water content is missing. This is due to a too strong cooling of the upper soil layers in the simulation, which inhibits melting in this layer. An episode with too early melting in all soil layers is observed at the end of the simulation period. As a consequence, the soil water content in all layers is too low after the melting compared to the observations.

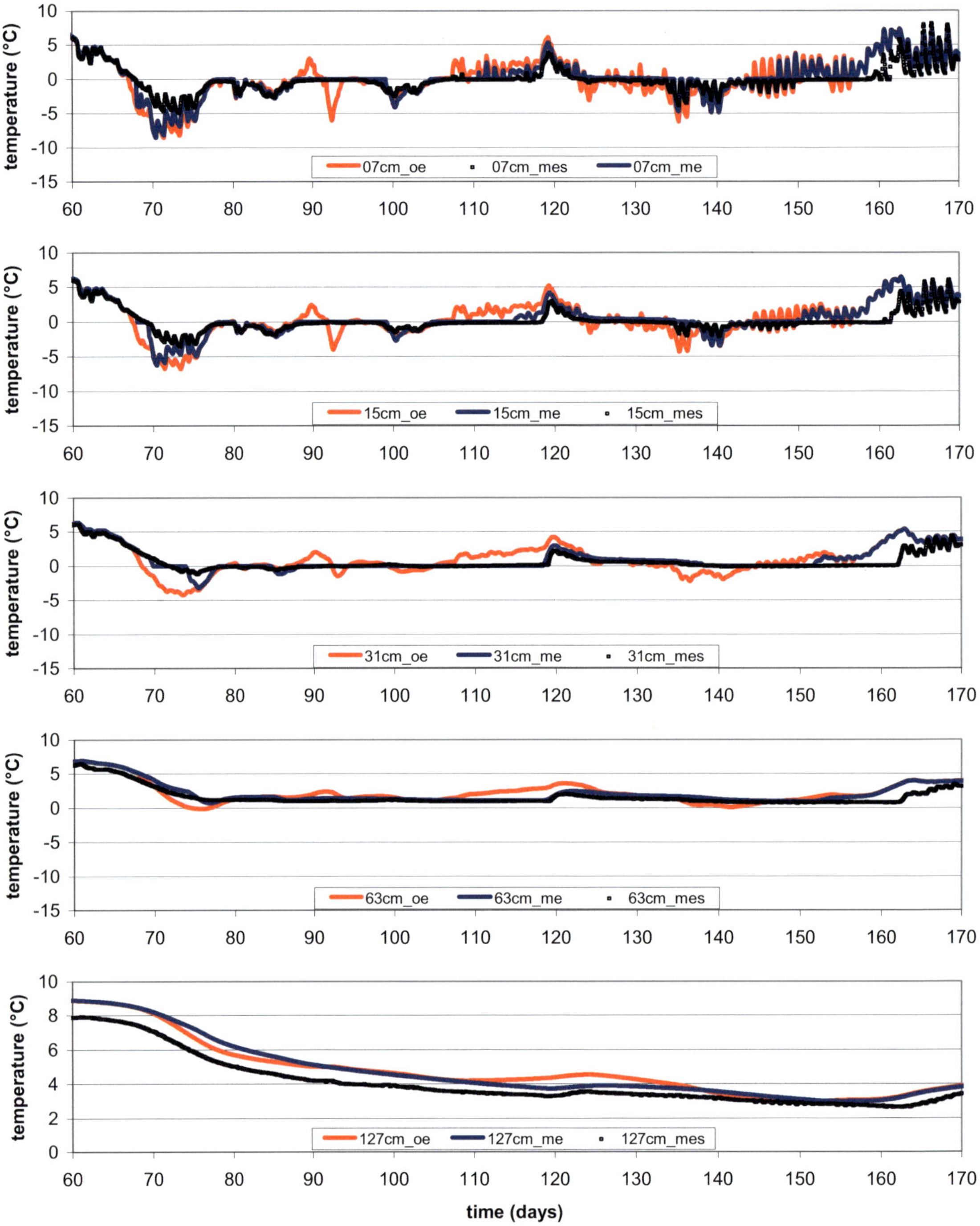

Figure 5.1: Observation (mes) and simulation of soil temperature for different soil depths with (me) and without (oe) soil freezing for the period from 1.10.2002 to 1.4.2003 at the Falkenberg measurement site.

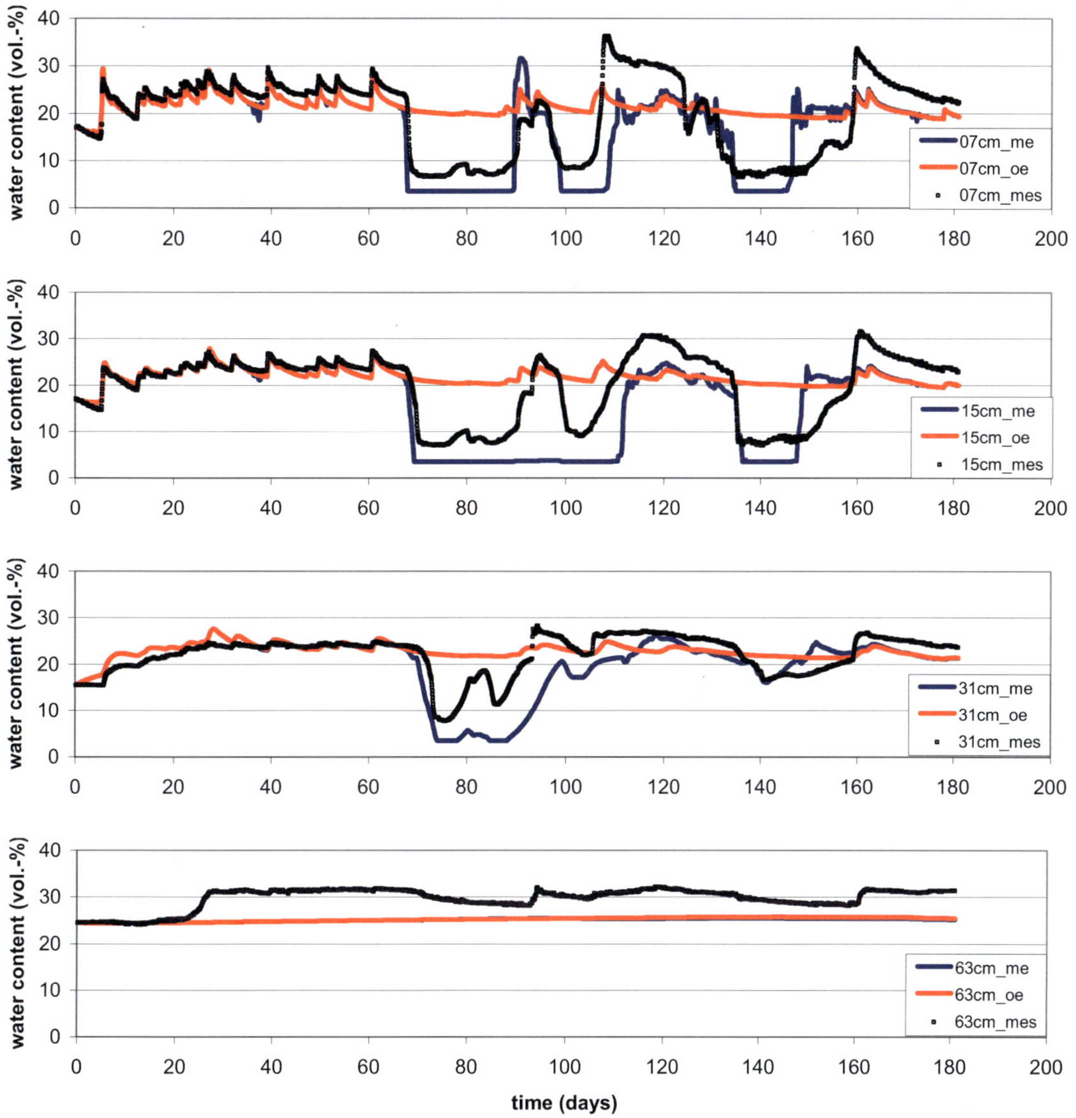

Figure 5.2: Observation (mes) and simulation of soil water content for different soil depths with (me) and without (oe) soil freezing for the period from 1.10.2002 to 1.4.2003 at the Falkenberg measurement site.

The soil water content of the simulation with and without soil freezing is similar at the end of the simulation period in all soil layers. This is in accordance to findings from CHERKAUER and LETTENMAIER (1999), who also found only a relatively small effect of frozen soil on soil moisture and runoff simulations. Larger differences in soil temperature profiles after melting between simulations with and without soil freezing can be observed, when the period of freezing reaches several days or weeks or when the depth of freezing reaches much deeper into the soil (LUO et al., 2002).

For the correct simulation of soil freezing, the thickness of soil layers is of minor importance. Using multiple layers, each with a thickness of 2 cm, yields similar results to the soil layer thickness given above.

The sensible heat flux is overestimated and the latent heat flux is underestimated in both simulations (Fig. 5.3). The difference in heat fluxes between the simulation with and without soil freezing is visible only for single days and does not exceed 40 W m^{-2}. On some days with soil freezing, the new parameterisation leads to an enhancement in simulation results and in some days it does not. This strongly depends on the snow cover on the soil. Longer periods of freezing should be considered for an evaluation of heat fluxes for frozen soil. The correlation coefficient is 0.58/0.57 for latent heat flux and 0.75/0.74 for sensible heat flux for the simulation with/without soil freezing. A strong underestimation of latent heat flux is mainly observed for melting periods where the observed soil water content is about 10 volume percent higher than the simulated soil water content.

For the simulation period from 1.6.2003 to 31.5.2004 we find episodes where soil freezing is simulated well with the new parameterisation. These are periods without snow cover on the soil or periods where the modelled snow height is in accordance to the observed one (Fig. 5.4). For those periods, the times of freezing and melting of soil water are simulated correctly and the soil temperature is simulated much better with the new parameterisation than without consideration of soil freezing. For episodes with snowfall, the performance of the model strongly depends on the modelled amount of snow.

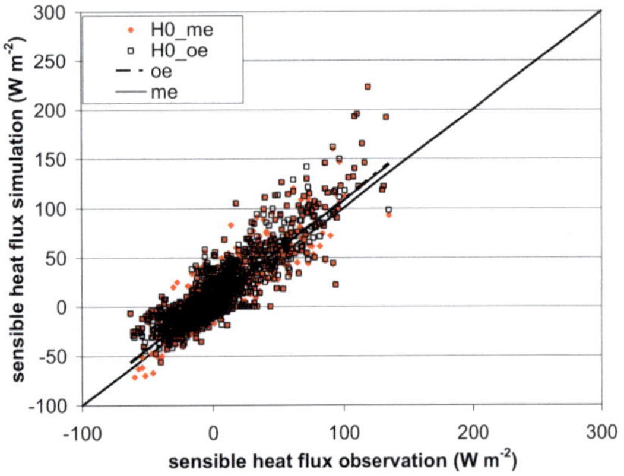

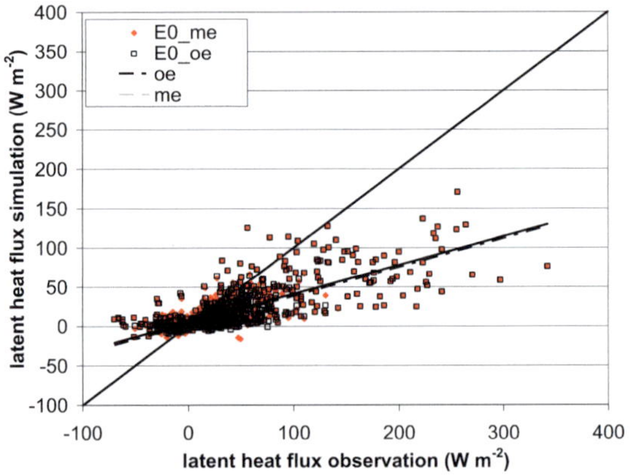

Figure 5.3: Scatter plot of sensible (left) and latent (right) heat fluxes for observation and simulation with (me) and without (oe) soil freezing for the period 1.12.2002 to 1.4.2003.

Since no snowfall observations are available, the model produces snowfall when the air temperature drops below the freezing point, which might produce too low snow amounts. The only observed value for a comparison is the measured snow height, which is measured once a day. When the model is not able to simulate a snow deck comparable to the observed one, the soil temperatures are simulated much too low and freezing of soil water starts too early (Fig. 5.5). Where no snow deck is observed and the model spuriously produces a snow deck, the melting of soil ice is delayed in the simulation (Fig. 5.5). From this simulation it becomes clear that the snow cover of the soil is a rather critical parameter for the correct simulation of soil temperature and soil moisture. This is in accordance to the findings of the PILPS 2(d) Experiment (LUO et al., 2002). VEG3D could only simulate correct snow coverage – allowing to conduct a reliable comparison of the new parameterisation for times with snow coverage – with hourly observed snowfall data and without any gaps in the observations.

For the periods of the ongoing SnowMIP2 experiment, such data is available and the model shows quite good results compared to other models and observations (SCHÄDLER (IMK), personal communication).

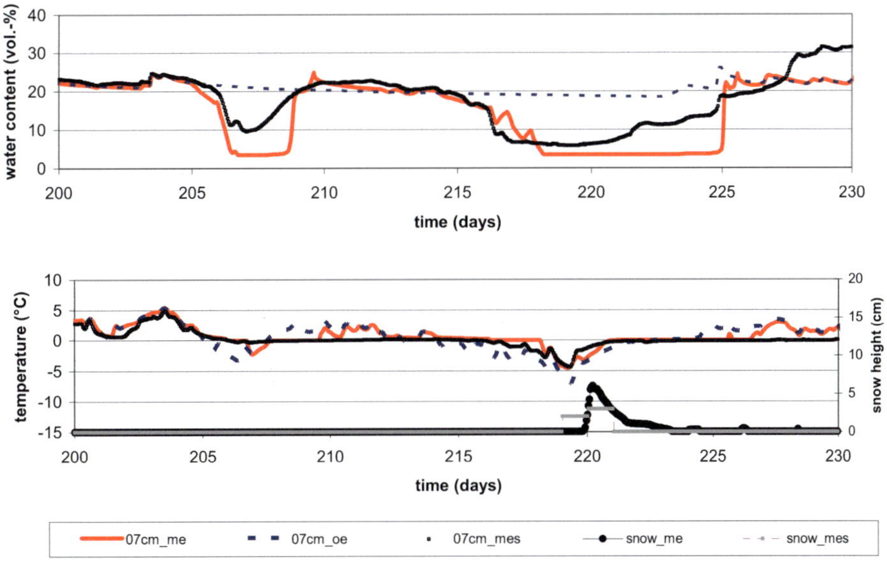

Figure 5.4: Winter period of the simulation for the Falkenberg measurement site from 1.6.2003 to 31.5.2004. Displayed is the water content of the simulations with (me) and without (oe) soil freezing and of the observations (mes) (top) and the soil temperature and the snow height (bottom).

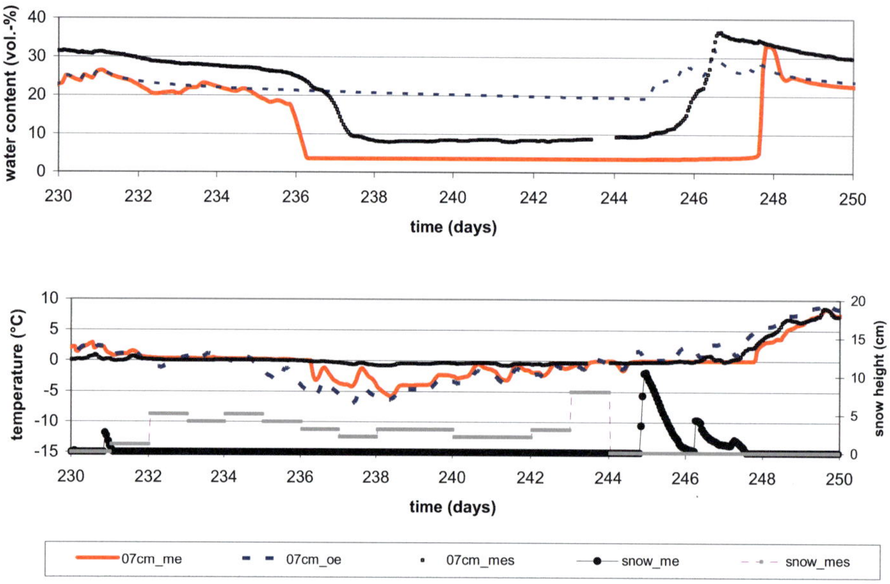

Figure 5.5: Winter period of the simulation for the Falkenberg measurement site from 1.6.2003 to 31.5.2004. Displayed is the water content of the simulations with (me) and without (oe) soil freezing and of the observations (top) and the soil temperature and the snow height (bottom).

5.2 VARIABLE SOIL TYPES WITHIN ONE SOIL COLUMN

Up to now, only one soil type within a single soil column has been considered in VEG3D. It is well known that the soil is stratified vertically in layers (horizons) of different soil types in nature and that the single layers may have significantly different soil characteristics that affect the soil water and soil heat transport markedly. Whether the introduction of soil horizons within one soil column would lead to a better prediction of soil water content and soil temperature was therefore tested in stand-alone simulations with VEG3D. These simulations show an improvement of the simulation results compared to measurements. Therefore, a digital soil map with vertically varying soil profiles for the area of Southwest Germany has been created based on the "Hydrologischer Atlas Deutschland" (HAD, 2003). This map can be used to prescribe the soil type in VEG3D simulations when the model is coupled online to the atmospheric part of COSMO-CLM.

5.2.1 Simulations with the VEG3D stand-alone model for vertically changing soil type

Soil analysis at the Falkenberg measurement site has shown that the topmost horizon down to a depth of 60 cm consists of sandy soil and a second horizon beneath it consists of loamy soil. Both soil types differ strongly in their soil hydraulic conductivity, which is high for sandy soils and low for loamy soils. Therefore, soil water content in the upper horizon is simulated too high if loamy soil is assumed for the whole soil column and it is simulated too low in the lower horizon if loamy sand is assumed for the whole soil column (Fig. 5.6). Assuming loamy sand in the uppermost five soil layers down to a depth of approximately 60 cm and loam in the lowermost three soil layers results in an improved simulation of soil water content for the period from 1.10.2002 to 1.4.2003 (Fig. 5.6). With this configuration we get realistic results in both horizons: the water content is kept at a lower level in the upper horizon and on a higher level in the lower horizon, agreeing more closely with observations. The advantage of the introduction of different soil horizons is only visible if the comparison is made for the whole soil column.

The simulation for the period from first of June 2003 to end of May 2004 gives similar results for vertical varying soil types (Fig. 5.7): a remarkable enhancement when using two soil horizons instead of loamy soil type can be noticed, and a clear enhancement for the lower soil layers when using vertical varying soil horizons instead of loamy sand in the whole column. Independent of soil type, frozen periods are well simulated. The increase in water content of the layer 60 cm deep in the late autumn is also visible for the period of 1.10.2002 to 1.4.2003 and is probably due to rising ground water. This increase is only simulated by the loamy sand soil type but the absolute soil water content is much lower than in the other two simulations and the observations.

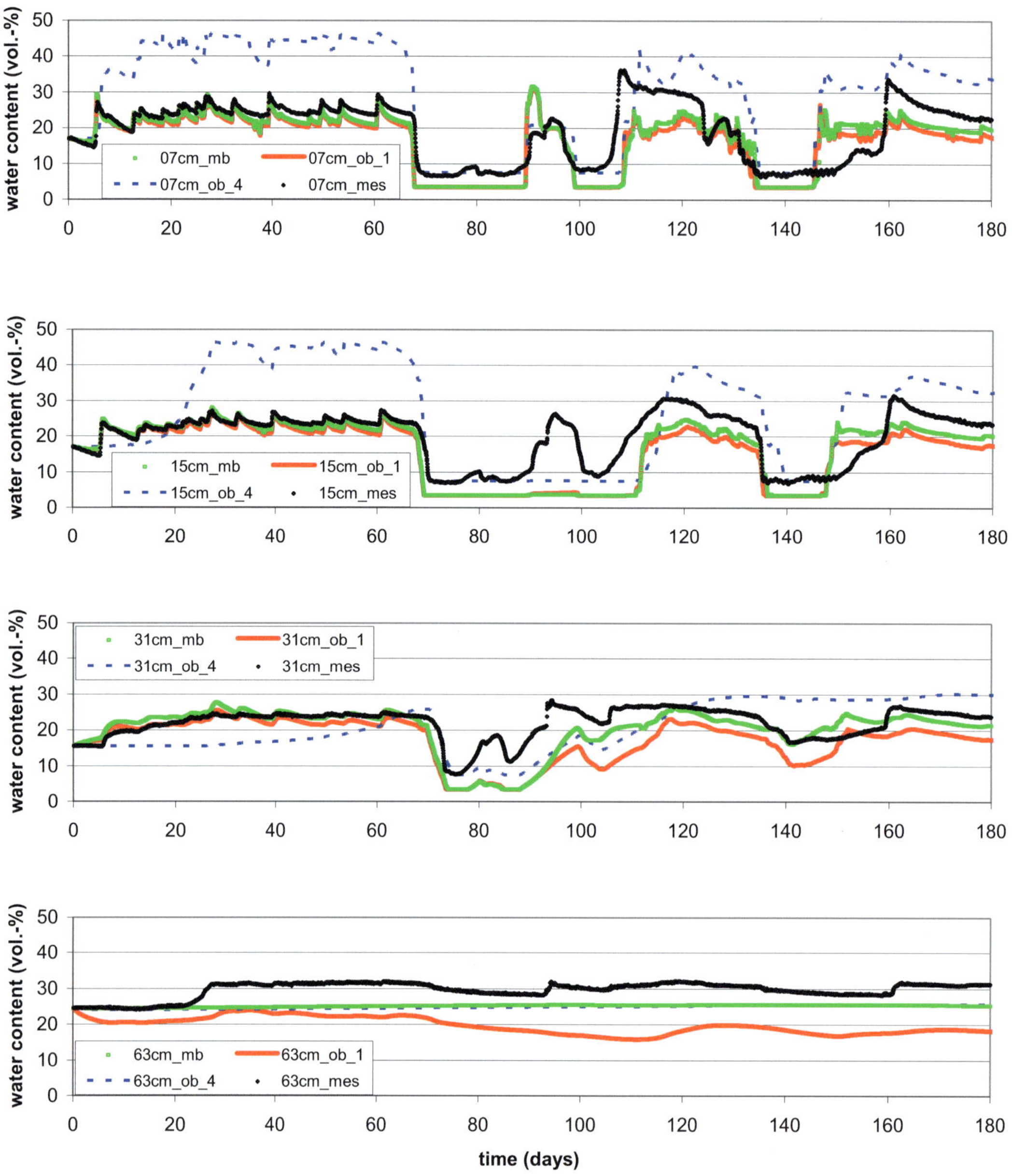

Figure 5.6: Observation (mes) and simulation of soil water content for different soil depths with (mb) and without (ob) varying soil type within the soil column for the period from 1.10.2002 to 1.4.2003 at Falkenberg measurement site. The simulations without varying soil type are depicted for loam (ob_4) and loamy sand (ob_1).

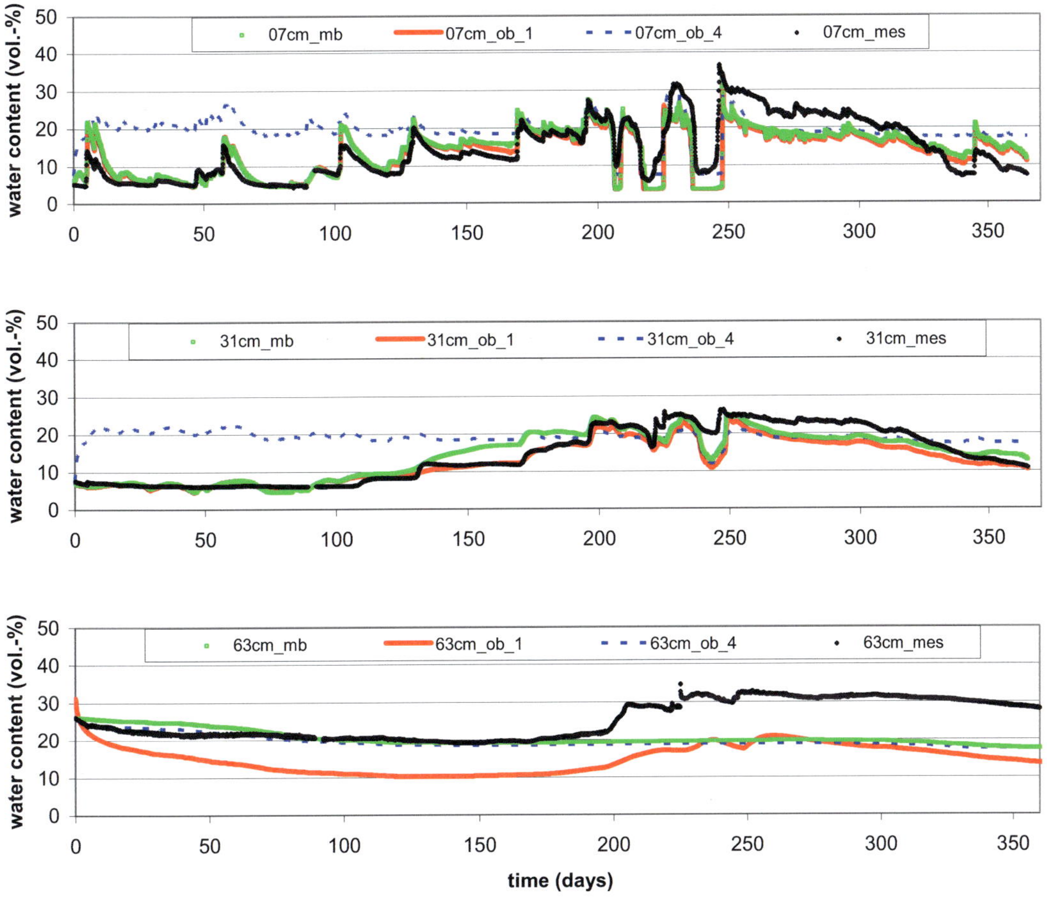

Figure 5.7: Observation (mes) and simulation of soil water content for different soil depths with (mb) and without (ob) varying soil type within the soil column for the period from 1.6.2003 to 31.5.2004 at Falkenberg measurement site. The simulations without varying soil type are depicted for loam (ob_4) and loamy sand (ob_1).

5.2.2 Soil type map for COSMO-CLM simulations with different soil types within one soil column

From the studies presented in the previous subsection it becomes obvious that vertical varying soil types lead to a better performance of the model. Therefore, VEG3D has been extended so that all soil characteristics are now available as three dimensional fields; they can also be initialised in this way. Information about the different soil horizons and their soil types in online coupled simulations with COSMO-CLM for the area of Southwest Germany is based on HAD. From a digital map of HAD the soil types are transferred in two steps (Fig. 5.8). Because not all soil types in HAD are available in VEG3D, we first appoint each of the 34 soil types in the HAD to one of the 16 soil types in VEG3D (this is detailed in Appendix 9.1). Secondly, the vertical partitioning of soil types is interpolated onto the VEG3D vertical grid. In this way, VEG3D can consider different soil horizons online coupled with COSMO-CLM. The difference in soil structure in the different soil layers is illustrated in Fig. 5.8. The parts outside Germany have been filled up by using a European soil type map with a resolution of 0.0625° and 15 different soil types. It would also be possible to choose the thickness of the soil layers of each individual grid box dependend on the information obtained from the HAD, but this feature has not yet been implemented and the depths of soil layers are the same for all grid boxes. To reveal the differences in soil type structure used in a COSMO-CLM simulation with the standard soil model TERRA_LM and in the simulation coupled with VEG3D, both soil maps are plotted for the first soil layer in Fig. 5.9. The corresponding soil types are given in Tab. 5.1. VEG3D considers 16 soil types and the TERRA_LM uses 10 different soil types. The VEG3D map is much more structured for Germany than the COSMO model standard map. The structure of the HAD is even more complex than the VEG3D map (Fig. 5.9).

The influence of different vertical soil types on the soil water budget has yet to be investigated for a coupled COSMO-CLM/VEG3D simulation. One problem is the lack of adequate observations, e.g. three dimensional soil moisture and soil temperature fields. These are necessary to evaluate whether the use of different soil types in one column improves the coupled simulations. The procedure developed has been modified by Wohnsiedler (2006) and some soil assignments from HAD to VEG3D have been changed.

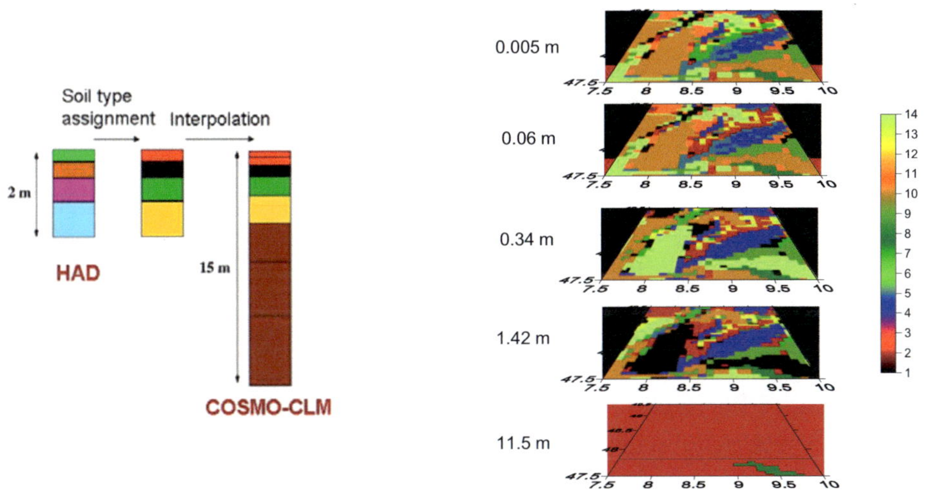

Figure 5.8: Process of soil type determination for VEG3D from HAD (left). Soil types in different soil layers in the COSMO model coupled with VEG3D obtained by the interpolation from HAD (right) for the Southwest of Germany. Soil type numbers are explained in Tab. 5.1.

Number	Soil type VEG3D	Number	Soil type TERRA_LM
1	loamy sand	1	Ice (17)
2	loam	2	rock (18)
3	silty clay loam	3	sand (9)
4	clay loam	4	sandy loam (7)
5	clay	5	loam (2)
6	peat	6	loamy clay (4)
7	sandy loam	7	clay (5)
8	water	8	peat (6)
9	sand	9	sea water (8)
10	silt loam	10	sea ice
11	sandy clay loam		
12	silt		
13	sandy clay		
14	silty clay		
17	ice		
18	rock		

Table 5.1: Soil types in VEG3D and TERRA_LM and their corresponding numbers. Numbers in brackets give the VEG3D soil types used when soil type is taken over from the DWD standard map for COSMO_CLM simulations coupled with VEG3D.

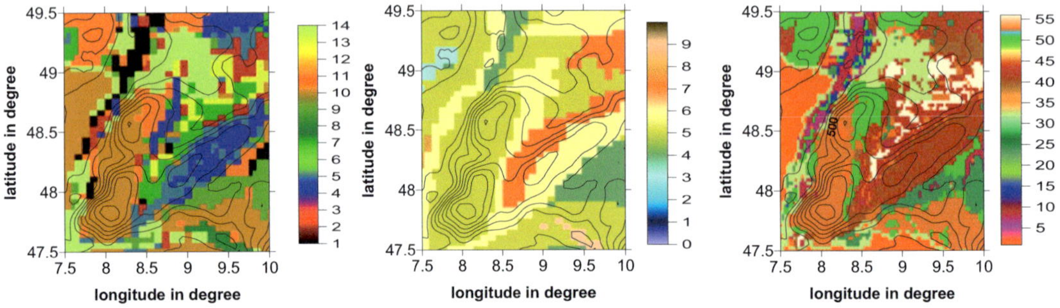

Figure 5.9: Soil types for the uppermost soil layer from the new vertical soil type map in VEG3D (left), from the TERRA_LM standard map (middle), and from the soil type classification of HAD (right). Soil type numbers of VEG3D and TERRA_LM are explained in Tab. 5.1. HAD classification is given in Appendix 9.1.

5.3 Implementation of VEG3D in COSMO-CLM

The online coupling of VEG3D with COSMO-CLM requires multiple adaptations and changes in the source code. In principle, the soil model TERRA_LM is replaced by the VEG3D and all routines, which use variables calculated by the soil-vegetation scheme need to be adapted to the variables used by VEG3D. Additional data sets needed for driving VEG3D (parameters, inventories) have to be implemented in the COSMO-CLM system.

5.3.1 New parameter data sets

VEG3D needs explicit land use information, which is not available in the external parameter data set provided by DWD. The DWD data set only contains the derived vegetation characteristics leaf area index and plant cover. In the pre-processor, these vegetation characteristics are computed as the area mean of the different land use classes occurring in one grid box, which are transferred from the CORINE, GLCC or GLC2000 land use data set (Ritter, personal communication, 2007). For VEG3D, the pre-processor calculates the majority land use class within one grid box as output; in an additional step, the 23 land use classes from the pre-processor are assigned to the 10 land use classes in VEG3D

(Tab. 5.2). The land use inventory thus produced (Fig. 5.10) is used as an input field during the setup of the simulation (Fig. 5.11).

Furthermore, the new three dimensional soil type fields described in subsection 5.2.2 can now be read in during this initialisation step. When no three dimensional soil map is given, the soil types from the external parameter data set from DWD are assigned to the soil types used in VEG3D (Tab. 5.1). The soil and vegetation characteristics for each soil type and land use class in VEG3D are included as a look-up table in the source code.

Up to now, ice and rock were not available as land use and soil types in VEG3D. In some alpine regions and in some Mediterranean areas ice or rock is used in the soil classification of the standard soil map. Therefore, ice and rock are now included in the soil data set of VEG3D. The new soil characteristics are given in Tab. 5.3. Ice has been introduced as an additional land use class in the land use data set. No vegetation cover is permitted for this class.

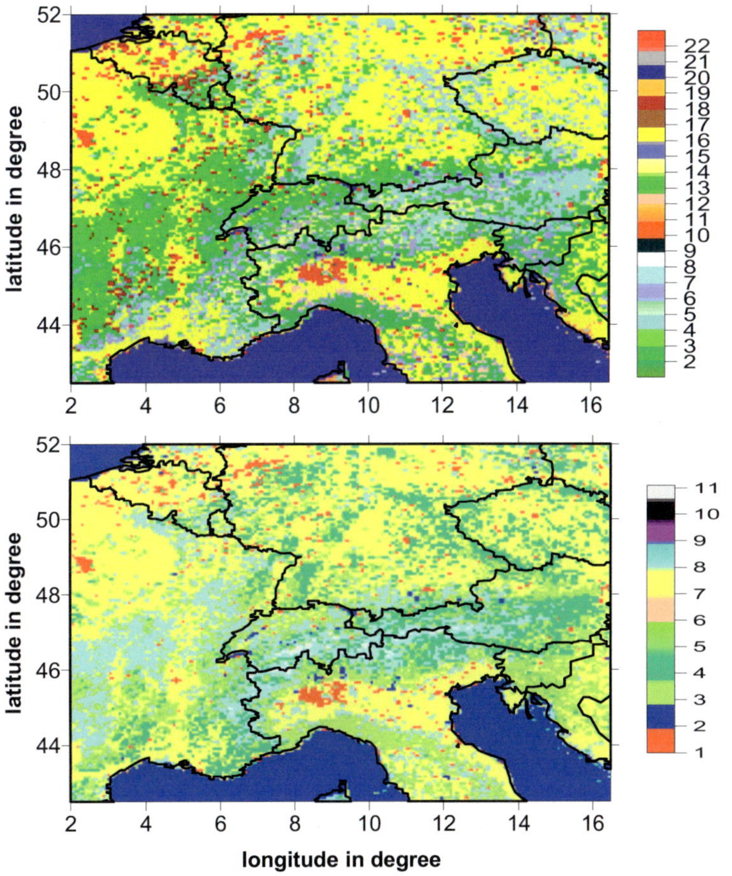

Figure 5.10: Land use classes of TERRA_LM (top) and VEG3D (bottom). Explanation of numbers is given in Tab. 5.2.

Number	Description VEG3D	Number	Description TERRA_LM
0	bare areas	*1*	evergreen broadleaf tree (3)
1	artificial surfaces	*2*	deciduous broadleaf tree closed (3)
2	water bodies	*3*	deciduous broadleaf tree open (3)
3	broadleaf tree	*4*	evergreen needleleaf trcc (4)
4	needleleaf tree	*5*	deciduous needleleaf tree (4)
5	mixed leaf tree	*6*	mixed leaf tree (5)
6	cultivated areas	*7*	fresh water flooded tree (3)
7	crop	*8*	saline water flooded tree (3)
8	grassland	*9*	mosaic tree / other nat. veg. (5)
9	mosaic grass / shrubs	*10*	burnt tree cover (5)
10	undefined	*11*	evergreen shrubs closed-open (8)
11	snow & ice	*12*	deciduous shrubs closed-open (8)
		13	herbaceous cover closed-open (8)
		14	sparse herbaceous or grass (8)
		15	flooded shrub or herbaceous (8)
		16	cultivated & managed areas (7)
		17	mosaic crop/tree/natural veg. (7)
		18	mosaic crop/shrub or grass (7)
		19	bare areas (0)
		20	water bodies (2)
		21	snow & ice (11)
		22	artificial surfaces (1)
		23	Undefined (1)

Table 5.2: Land use classes in VEG3D and TERRA_LM and their associated numbers. Numbers in brackets give the VEG3D land use type assigned to the TERRA_LM land use type.

Parameter	Variable name	Rock	Ice	Unit
Saturation water content	dws	0.2	0.2	-
Residual soil water content	dwr	0.05	0.05	-
Matric potential at saturation	dpsis	-1	-1	m
Hydraulic conductivity at saturation	dks	0	0	m/s
Parameter for soil hydrology	dlamb	1	1	-
Albedo of dry soil	dg0	0.15	0.3	-
Albedo difference dry - saturated soil	dgl	0	0	-
Emissivity of dry soil	deg0	0.98	0.9	-
Emis. difference of dry and wet soil	degl	0	0	-
Quartz content	dqu	0.5	0.5	-
Heat capacity	dcap	2.10E+06	1.80E+06	J/m^3K
Density of dry soil	drhod	2700	900	kg/m^3
Thermal conductivity of dry soil	dcondd	2.93	2.51	W/mK
Thermal conductivity of saturated soil	dconds	2.93	2.51	W/mK

Table 5.3: New parameter data set in VEG3D for ice and rock.

5.3.2 Coupling of VEG3D with the atmospheric part of the model

All variables that are output variables in the standard soil model of COSMO-CLM and are used in other parts of the COSMO-CLM have to be calculated from variables available in VEG3D. This assignment is done at the end of the VEG3D time step.

All parameters used in VEG3D that are not defined as global parameters have to be prescribed new in the VEG3D routine at each time step in order to be available even after restart, e.g. time-dependent vegetation parameters like leaf area index or plant cover.

It is important to write the variables that are only used in the VEG3D routine into the restart file in order to enable a correct restart procedure after a certain simulation time. These include the two snow temperatures, the heat fluxes, the vegetation temperature, the canopy temperature, and specific humidity.

The soil-vegetation model has an explicit influence on three physics routines and the diagnostic routine within the source code where variables from the soil-vegetation model are used (Fig. 5.11). Therefore, the following routines have to be adapted to the soil model used:

a) Radiation routine

The surface temperature and the albedo from the soil-vegetation model are used for the calculation of the thermal infrared and reflected shortwave radiation from the ground. The surface temperature is needed for the calculation of thermal radiation from the surface and normally the snow weighted soil temperature from TERRA_LM is taken; when coupled with VEG3D, the canopy temperature is taken instead.

The albedo is needed for the calculation of the shortwave radiation from the ground. TERRA_LM calculates a mean albedo from the albedo of uncovered soil, of vegetation, and of snow. The albedo of vegetation is kept constant in TERRA_LM, whereas VEG3D calculates a time-dependent vegetation albedo, which is used in the radiation routine for the summed albedo at the surface together with the soil and the snow albedo from VEG3D.

Operationally, only the net radiation balance is available from the radiation routine. However, VEG3D needs the incoming short and long wave radiation, which has to be stored as additional output parameters.

b) Convection routine

Deciding whether snow or rain reaches the ground involves considering the temperature at 300 m above the surface and the surface temperature (TERRA_LM) or the canopy temperature (VEG3D).

c) Turbulence routine

As lower values for the calculation of the surface fluxes by a bulk scheme, the surface humidity and the surface temperature are taken when using TERRA_LM. Using VEG3D, the heat fluxes are already calculated by the soil-vegetation scheme and are readily available for use in this routine.

d) Diagnostics

At the end of each time step, a diagnostic routine is called to calculate near-surface values, e.g. 2m-temperature, 2m-humidity and near-surface winds. Using VEG3D, this part is switched off and these parameters are calculated directly in the soil-vegetation model.

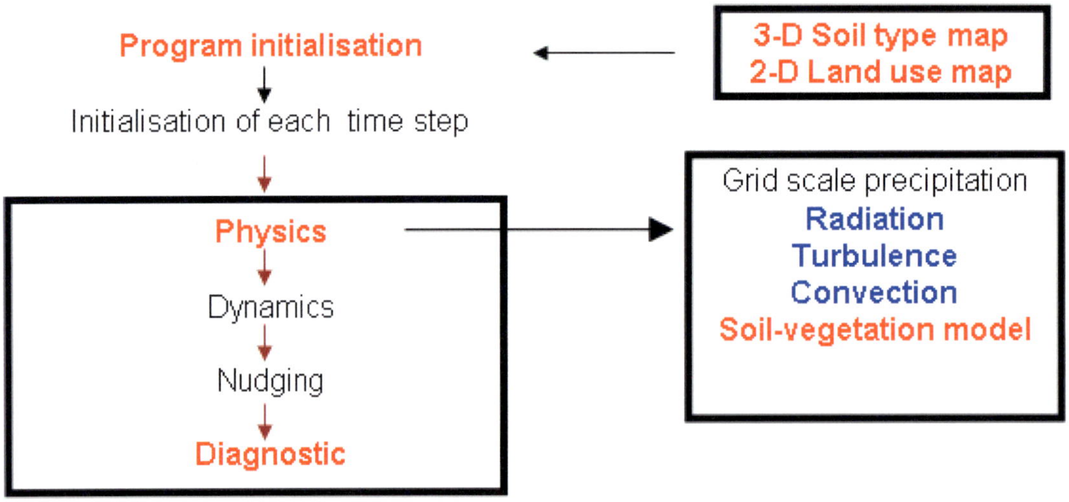

Figure 5.11: Sketch of the simulation steps within the COSMO-CLM. In red: major changes in the source code that apply when the VEG3D routine is used. In blue: routines where variables from the soil vegetation model are used.

6 Sensitivity of COSMO-CLM regional climate simulations with respect to VEG3D and TERRA_LM land surface scheme

In numerical weather forecast models, the detailed description of soil processes was neglected for a long time. Up to now, soil moisture has been used as a tuning parameter to adjust the 2m-temperature by variational soil moisture analysis (Hess, 2001, Rhodin et al., 1999) in the operational runs of the COSMO model from DWD. In the 1990s, the importance of a correct description of the soil processes was recognized and intensive studies were carried out on the influence of land surface schemes on precipitation (e.g. Timbal and Henderson-Sellers, 1998), the partitioning of surface heat fluxes (e.g. Pitman and Henderson-Sellers, 1998) and the development of boundary layer and convection (e.g. Beljaars et al., 1996). The two-layer force-restore method schemes for temperature (Deardorff, 1978) and bucket models for soil water content (Manabe, 1969) used in the past have been replaced by complex multilayer soil models and even soil moisture freezing and different soil horizons are taken into account. Furthermore, in some models the subgrid scale heterogeneity of the land surface is modelled by so-called tile approaches (Avissar and Pielke, 1989). Because soil processes are slow processes at the lower boundary of the atmosphere, their detailed description at all scales is even more important in climate models than in weather prediction models. A spurious simulation of soil water content can lead to biases in temperature and precipitation (Vidale et al., 2003). The soil water content is of crucial importance especially for the amount of convective precipitation (Schär et al., 1999).

The description of soil processes is very complex and many different soil-vegetation-atmosphere transfer schemes (SVATs) are used in regional climate models, e.g. BATS (Dickinson et al., 1993), ISBA (Noilhan and Mahfouf, 1996) and MOSES (Cox et al., 1999). Besides the different formulations for soil water and soil heat transport they also differ considerably in the parameterisation of vegetation. In general, there are two possibilities for considering vegetation in land surface models. The first category of schemes does not consider an explicit vegetation layer (e.g. COSMO-CLM (Will et al., 2007),

REMO (JACOB, 2001), HIRHAM (CHRISTENSEN et al., 1996)). They assume that the vegetation has the same temperature as the soil and that the fluxes from the canopy can be described by quantities already available in the soil model or by parameterisation and vegetation parameters. The surface temperature is used to calculate the sensible and latent heat exchange between surface and atmosphere. The second category explicitly considers at least one vegetation layer (e.g. RegCM (GIORGI et al., 1993), MM5 (CHEN and DUDHIA, 2001)) and the schemes calculate transpiration and heat flux from the foliage to the canopy air and from the canopy air to the lower atmosphere. Since the soil-vegetation scheme influences the atmospheric processes via the exchange of latent and sensible heat, there is the question of how quantities like near-surface temperature and precipitation are affected by the use of different land surface schemes in high-resolution climate simulations for complex terrain.

To study this, the VEG3D SVAT scheme was implemented in the regional climate model COSMO-CLM and simulations with a 7 km horizontal resolution were performed for Southwest Germany. VEG3D has an explicit vegetation layer, in contrast to the operationally used soil model TERRA_LM (DOMS et al., 2005), which has no explicit soil-vegetation layer. One therefore expects to find considerable differences between the two schemes, especially over high vegetation. For detailed comparisons, simulations with the stand-alone versions of the two SVAT schemes were performed. Such simulations allow study of the effect of the land surface schemes on soil water content, soil temperature and heat fluxes without any atmospheric feedback. Simulations extending over several years were performed to assess the influence of the land surface scheme on the COSMO-CLM simulations, with each SVAT coupled online with the atmospheric part of the model.

Section 6.1 describes briefly the two soil models used and highlights the differences between them. Section 6.2 presents a comparison of the stand-alone versions of both models driven by observations for low and high vegetation. COSMO-CLM simulations with each of the two soil models coupled online are compared for a single day in section 6.3 and for the year 2001 in section 6.4. Section 6.5 presents the comparison of the COSMO-CLM simulations with each of the two soil models coupled online for the period 1991 to 1995 and the decade from 1991 to 2000. A brief summary is given in section 6.6.

6.1 Description of the Models

6.1.1 TERRA_LM

TERRA_LM is a multi-layer soil model without an explicit vegetation layer. It distinguishes five different soil types and also considers ice, rock and water. A changing vertical profile of soil temperature and soil water content is only calculated for these five soil types. In case of ice, the water content is kept constant and potential evaporation from an ice surface is assumed. For water, the surface temperature is taken from the driving data: for rock, no water transport in the soil is considered, but a soil temperature profile is calculated. The vegetation parameters like leaf area index and fraction of plant cover are calculated as the weighted area mean of the different land use classes occurring in the grid box considered. In climate mode the model usually runs with 10 vertical soil layers with the following depths of main levels (m): 0.005, 0.025, 0.07, 0.16, 0.34, 0.7, 1.42, 2.86, 6.74, 11.5.

In the lowest soil layer, temperature is kept constant at a climatological value, which is determined by the climatological mean of near-surface temperature. There are active and passive hydrological soil layers in TERRA_LM; the latter are the deeper soil layers. In the active layers, water transport is calculated explicitly and below these active layers, water content is kept at the value of the lowest hydrological active layer for thermal calculations in the passive layers and no water transport is calculated. Gravitational flow is assumed as a lower boundary condition for the lowest active hydrological layer. In the standard COSMO model, the hydrological active layers reach down to 2.50 m.

The model has an interception store and a snow store where rain or snow can be accumulated. If one of these stores reaches its maximum capacity, the surplus can infiltrate into the soil. The model produces surface runoff if the infiltration rate is too low for the precipitation rate or for the surplus from the interception store. When the soil water content reaches the field capacity, ground runoff from the soil is generated and the water is taken out of the modelling system. There is no lateral water flow in the soil.

In case of snowfall, the model produces a snow cover on the soil and from a height of 0.01 m onward it is considered as a snow deck and a snow temperature is calculated implicitly. In TERRA_LM it is possible that a grid cell is covered partially by snow and par tially not. The snow temperature is calculated from a prognostic equation and the snow surface temperature, which is an output variable, is calculated diagnostically by linear extrapolation. Snow density depends on the snow depth and also on its temperature. Ageing of snow results in a decrease of albedo and an increase of density. The thermal conductivity of snow is parameterised depending on the amount of snow water.

The heat fluxes from the surface are calculated by a bulk formulation using the temperature and the specific humidity at the surface calculated by the soil model and the values in the lowest atmospheric level. The surface humidity is calculated from the sum of evaporation from snow store, from interception store, from bare soil, transpiration from all soil layers and formation of dew and rime. The surface temperature is a weighted mean between the temperature of the uppermost soil layer and the snow temperature. The calculation of these fluxes is part of the turbulence parameterisation (DOMS et al., 2005).

Bare soil evaporation is calculated as the minimum of potential evaporation at surface temperature and the maximum moisture flux through the surface the model can sustain (DICKINSON, 1984). Plant transpiration Tr is parameterised basically following DICKINSON (1984) as a function of potential evaporation E_{pot} at surface temperature T_{sfc}, the atmospheric resistance r_a and the foliage resistance r_f:

$$Tr = r_a (r_a + r_f)^{-1} E_{pot}(T_{sfc}). \qquad (6.1)$$

In this formulation it is assumed that the foliage temperature equals the surface temperature and that the moisture flux between the foliage and the air inside the canopy equals the flux between the air inside and the air above the canopy. The foliage resistance is a function of radiation, soil water content in the root zone, 2m-temperature, leaf area index and the maximum and minimum values of stomatal resistance. The atmospheric resistance is the product of the bulk transfer coefficient for moisture and the magnitude of the wind in the first model layer above the surface.

The transport of liquid water w_l in the soil is calculated by solving the prognostic Richards equation (HILLEL, 1980):

$$\frac{\partial w_l(z,t)}{\partial t} = \frac{\partial}{\partial z}\left[D_w(w_l,z)\frac{\partial w_l(z,t)}{\partial z} - K_w(w_l,z)\right] - r_w(w_l,z). \qquad (6.2)$$

The water uptake by plants from the soil is described by r_w. In this formulation, hydraulic diffusivity D_w and hydraulic conductivity K_w have to be parameterised, which is done using exponential functions by RIJTEMA (1969):

$$D_w(\overline{w_l}) = D_0 \exp\left[D_1(w_{PV} - \overline{w_l})/(w_{PV} - w_{ADP})\right], \quad (6.3)$$

$$K_w(\overline{w_l}) = K_0 \exp\left[K_1(w_{PV} - \overline{w_l})/(w_{PV} - w_{ADP}\right], \quad (6.4)$$

where D_0, D_1, K_0, and K_1 are hydraulic parameters depednig on soil type, w_{PV} and w_{ADP} are the pore volume and the air dryness point and $\overline{w_l}$ is the weighted mean of the liquid water content at half levels. In the presence of ice

$$w_l' = \frac{w_l}{1 - w_{ice}} \quad (6.5)$$

replaces the water content w_l and the hydraulic coefficients are reduced by the reduction factor

$$r = 1 - \frac{Max(w_{ice,k}, w_{ice,k+1})}{w_{PV}}. \quad (6.6)$$

The reduction factor depends on the ice content w_{ice} of the two adjacent soil layers k and $k+1$ and the pore volume w_{PV}.

The change in soil temperature T_B is described by the molecular heat equation:

$$\frac{\partial T_B}{\partial t} = \frac{1}{\rho c}\frac{\partial}{\partial z}\left(\lambda_B \frac{\partial T_B}{\partial z}\right). \quad (6.7)$$

The volumetric heat capacity of the soil ρc is the sum of the volumetric heat capacity of the soil components dry soil ($\rho_o c_o$), water content ($\rho_l c_l$), and ice content ($\rho_{ice} c_{ice}$):

$$\rho c = \rho_o c_0 (1 - w_l - w_{ice}) + \rho_w c_w w_l + \rho_w c_{ice} w_{ice}. \quad (6.8)$$

The heat conductivity λ_B is parameterised with a constant soil water content depending only on soil type (calculated as the mean of soil water content at field capacity and of soil water content at plant wilting point) and does not consider soil ice. Due to the very thin uppermost layer, the soil heat equation has to be solved implicitly. The change of soil ice, soil water, and soil temperature caused by freezing and melting processes in the soil is calculated diagnostically at the end of each time step.

The vegetation is accounted for by the leaf area index and the plant cover, which vary from minimum to maximum values during the vegetation period depending on Julian day, latitude and altitude, by the root depth depending on the Julian day and by the time independent roughness length; soil horizons cannot be modelled.

Near-surface variables, e.g. 2m-temperature and wind speed in 10 m height, are calculated diagnostically by using a scheme based on similarity theory and developed by RASCHENDORFER (DWD):

$$\phi_L = \phi_S + \frac{f_{SI}}{f_{SA}}(\phi_A - \phi_S). \tag{6.9}$$

The parameter ϕ_S is the variable value at the surface, ϕ_A is the variable value at the lowermost atmospheric layer and ϕ_L is the value at the height considered. The parameters f_{SI} and f_{IA} are dimensionless resistance parameters describing the resistance for the layer from the surface level S to the considered level L and from the level L to the lowermost atmospheric level A.

6.1.2 VEG3D

VEG3D is a SVAT model with an explicit vegetation layer (SCHÄDLER, 1990). The vegetation is represented by the "big leaf" concept, where the vegetation is thought of as a single layer above the soil exchanging latent and sensible heat with the canopy air and extends DEARDORFF's (1978) model. The vegetation layer is considered massless and has its own vegetation temperature. The model distinguishes 13 different soil types. Additionally it accounts for ice, rock, and water, which are handled in a similar way to TERRA_LM. Unlike TERRA_LM, VEG3D uses only the majority land use class within one grid box to determine the vegetation characteristics for the grid box. The vertical partition of soil layers has been adjusted to be the same as in TERRA_LM. At the lowest soil layer, the temperature is kept constant and gravitational flux is used as boundary condition for soil moisture like in TERRA_LM. Unlike TERRA_LM, there is no distinction between active and passive hydrological layers. All layers are active layers. The model has an interception and a snow store and produces surface runoff if the infiltration rate is too low for the precipitation rate or the surplus in the interception store. Runoff from soil layers is generated if the soil water content becomes equal to the pore volume. In contrast to TERRA_LM, the runoff cannot leave the soil but is transported to the layers above. Runoff is only produced as surface runoff when all layers are saturated. Subsurface runoff, therefore, can only be generated by gravitational transport from the lowest soil layer.

In case of snowfall, the model produces a snow cover on the soil. From a height of 0.01 m onward it is considered as a snow deck, for which a snow temperature is calculated, as in TERRA_LM. In VEG3D, it is not possible for a grid cell to be covered partially by snow. If a snow deck is simulated, all the ground not covered by vegetation is assumed to be covered by snow. The snow on the ground is modelled as one layer, for which an individual prognostic snow temperature at the top and the bottom of the layer and a snow density is calculated. The snow model is based on the Canadian Land Surface Scheme (CLASS) (VERSEGHY, 1991) and the snow model ISBA1 (DOUVILLE et al., 1995). The albedo, the emissivity and the transfer coefficients for the heat fluxes of the vegetation are adjusted for snow-covered vegetation.

The heat fluxes from the surface are calculated by a bulk formula that takes account of the temperature and the specific humidity within the canopy and the values in the lowest atmospheric level. The transfer coefficients are calculated according to similarity theory. For the closure, the Monin-Obukhov length is calculated iteratively. The values within the canopy are calculated by considering the fluxes from bare soil and foliage. To calculate the fluxes from bare soil, the surface humidity is interpolated between the saturation humidity of the first soil layer and the humidity within the canopy. Either the temperature of the uppermost soil layer or the snow temperature is taken as surface temperature. Neutral conditions and logarithmical vertical profiles based on displacement height are assumed to calculate the transfer coefficients within the canopy.

Plant transpiration is expressed as a function of saturation humidity calculated from the foliage temperature T_f and humidity of the first model level q_{af}:

$$Tr = (r_s + r_f)^{-1}(q_s(T_f) - q_{af}), \tag{6.10}$$

where r_f is the resistance from the foliage to the surrounding air calculated by empirical functions according to TACONET et al. (1986) and r_s is the stomatal resistance after TACONET et al. (1986) and DEARDORFF (1978). When the interception store of the plant surface is not zero, potential evaporation from the interception store is calculated. The temperature of the foliage is determined iteratively from the energy balance of the leaves.

The Richards equation for water transport is used in VEG3D in its potential form:

$$C_w(\psi, z)\frac{\partial \psi(z,t)}{\partial t} = \frac{\partial}{\partial z}\left[K(\psi, z)\left(\frac{\partial \psi(z,t)}{\partial z} - 1\right)\right] - r_w(w_l, z), \qquad (6.11)$$

$$C_w = \frac{\partial w_l}{\partial \psi}, \; K = D\frac{\partial w_l}{\partial \psi}.$$

To calculate the matric potential ψ and the hydraulic conductivity K, three different parameterisations are available in line with BROOKS and COREY (1964), CAMPBELL (1974) and VAN GENUCHTEN (1980). In most cases, the Van Genuchten parameterisation, which is often used in hydrological models, shows the best results in soil moisture simulation (BRAUN and SCHÄDLER, 2005). It is therefore used for the following simulations and is already described in detail together with freezing and melting of soil in subsection 5.1.2.

As in TERRA_LM, the soil temperature is calculated by the molecular heat transport equation. In contrast to TERRA_LM, the thermal conductivity is a function of the actual water content and is changed in the presence of ice. The heat capacity is calculated as it is in TERRA_LM.

The vegetation is described by the parameters leaf area index, plant cover, roughness length and displacement height, which vary from minimum to maximum values during the vegetation period depending on Julian day and the time-independent root depth, which is determined by the vegetation type. Different soil horizons can be assigned within one soil column, a feature not used in this study in order to provide a fair comparison with TERRA_LM. For the same reason, only the five soil types in VEG3D that are available in TERRA_LM are used.

The near-surface parameters, e.g. 2m-temperature and 10m-wind speed, are calculated from the integral form of the flux gradient relationship. For 2m-temperature, the following form is used:

$$\theta_{2m} = \frac{\theta_*}{k}\left[f_T\left(\frac{z_{oh}+2}{L}\right) - f_T\left(\frac{z_r - d}{L}\right) + \theta_a \right], \tag{6.12}$$

where L is the Monin-Obukhov length, θ_a is the atmospheric potential temperature at the lowest atmospheric model level with height z_r, θ_* is the generic turbulent temperature scale, z_{oh} is the roughness length for heat, d is the displacement height and f_T is the integrated Dyer-Businger relationship for heat.

The main differences between the two SVAT models are summarized in Tab. 6.1.

Process	**TERRA_LM**	**VEG3D**
Runoff	Oversaturation in soil layers goes directly into runoff	Oversaturation is transported to soil layer above and can leave model only as surface runoff
Parameterisation of soil water transport	Rijtema (1969)	Van Genuchten (1980)
Snow	Partial snow cover possible	Snow deck only for whole grid box
Thermal conductivity of the soil	Constant soil water content used; no change in the presence of ice	Actual soil water content use; changed in the presence of ice
Plant characteristics	Mean value over grid box	Only major plant form considered
Vegetation	No explicit vegetation layer	Big leaf concept for vegetation
Surface fluxes	Calculation by surface values	Calculation by values inside the canopy

Table 6.1: Main differences between the land surface models TERRA_LM and VEG3D.

6.2 SIMULATIONS WITH THE STAND-ALONE VERSIONS OF THE TWO SOIL-VEGETATION MODELS

A suitable way of comparing two SVAT models is to drive them with observed meteorological data in a stand-alone mode without coupling to an atmospheric model. With the same forcing data, it is possible to determine the differences in soil temperature and soil moisture as well as in heat fluxes that occur due to the different model formulations, without possible feedback sensitivities of the atmospheric model.

Due to the different model formulations, differences that are particularly large over high vegetation would be expected. To compare the capability of both models to simulate low and high vegetation, simulations for the Falkenberg site (NEISSER et al., 2002) and the Hartheim site (MAYER et al., 2000) were performed (Fig. 6.1). The measurement field Falkenberg operated by DWD's Observatorium Lindenberg is an open grassland site southeast of Berlin; Hartheim is an experimental forest site, including an observation tower, situated in a pine forest 1 km east of the river Rhine, operated by the University of Freiburg. The year 2005 was simulated for Lindenberg and the year 2001 for Hartheim.

Driving the soil models with observations requires the option to calculate the heat fluxes at the surface within the stand-alone model. In TERRA_LM this is normally part of a turbulence scheme in the atmospheric part of the COSMO-CLM, which therefore has to be included in the stand-alone version. The included scheme is based on LOUIS (1979) and was formerly used in the COSMO model. In VEG3D the heat fluxes are already calculated according to similarity theory by the soil–vegetation scheme.

Besides atmospheric driving data, the plant and soil parameters of the two measurement sites are needed. There are two ways to determine these characteristics: the soil type and the land use can be determined at the observation site and a parameter table can be used to read off the soil and vegetation parameters or the characteristics used by COSMO-CLM by default at this grid point are taken. Here, the first option was chosen and the parameters were taken from the tables given in the COSMO model documentation (DOMS et al., 2005) for TERRA_LM and from the tables contained in VEG3D.

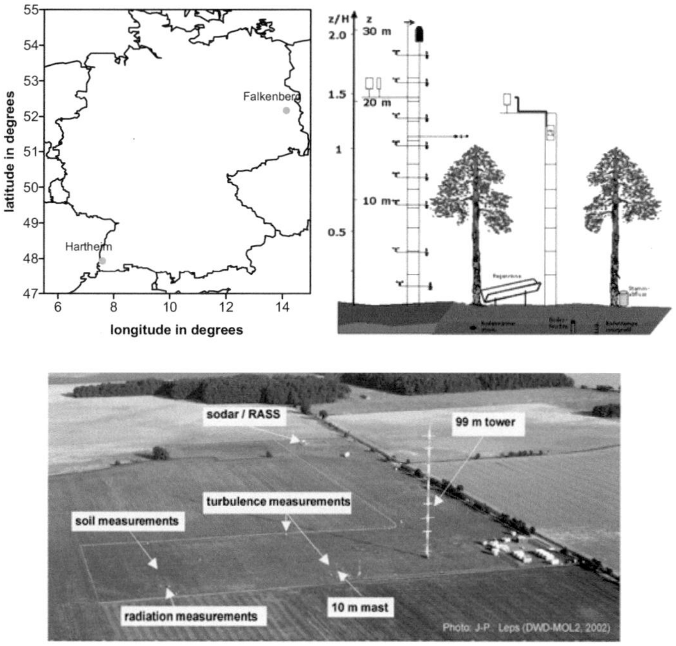

Figure 6.1: Map of the observation sites (upper left). Stations are indicated by grey points. Sketch of the observational tower in Hartheim (upper right, courtesy of the Meteorologisches Institut der Universität Freiburg) and picture of the measurement field Falkenberg (bottom).

Differences between these two approaches can arise because in COSMO-CLM the vegetation characteristics and the soil type are an average over the land use and soil classes classified in a grid box, so they can differ markedly from the observed ones at the observation site. This could affect the simulation results considerably because the model results are quite sensitive to changes in plant and soil characteristics. Because of this sensitivity, a special data set of leaf area index, fraction of plant cover and soil characteristics has been collected by HERET et al. (2006) for Lindenberg, which is used in this comparison for the TERRA_LM simulation. The vegetation characteristics, soil types and measurement heights used for each simulation are shown in Tab. 6.2 and Tab. 6.3. In all stand-alone simulations only the lowest soil layer in TERRA_LM is a hydrological passive layer. All the other layers are hydrological active layers.

The models were initialised with the same soil temperature and soil water content profiles on January 1 of each year and were driven by the hourly means of meteorological variables. The models were run with 10 soil layers with main level depths as set out in section 6.2 and the model output was every hour.

Plant/soil characteristics	Lindenberg TERRA_LM	Lindenberg VEG3D	Hartheim TERRA_LM	Hartheim VEG3D
Roughness length (m)	0.03	Minimum: 0.02 Maximum: 0.03	1.0	1.0
Displacement height (m)	--	Minimum: 0.1 Maximum:0.15	--	12.0
Leaf area index minimum	0.5	2.0	1.3	9.0
Leaf area index maximum	2.5	4.0	3.8	13.0
Plant cover minimum	0.55	0.90	0.8	0.9
Plant cover maximum	0.8	0.95	0.8	0.9
Root depth (m)	0.6	1.55	0.6	2.0
Land use	grassland	grassland	coniferous forest	coniferous forest
Soil type	loamy sand	loamy sand	sandy loam	sandy loam

Table 6.2: Plant and soil characteristics of TERRA_LM and VEG3D at Lindenberg (grassland) and Hartheim (coniferous forest).

Parameter	Lindenberg	Hartheim
Temperature	2 m	19.1 m
Humidity	2 m	19.1 m
Wind speed	10 m	18.8 m
Radiation	2 m	16 m
Precipitation	Surface	Plant free area
Soil temperature (depth in m)	0.05, 0.1, 0.15, 0.2, 0.3, 0.45, 0.5, 0.6, 0.9, 1.0, 1.2	0.01, 0.03, 0.05, 0.1, 0.2, 0.4
Average soil moisture content (depth in m)	0.03-0.09, 0.09-0.27, 0.27-0.81	Not available
Sensible and latent heat flux	2 m	Calculated from temperature and humidity profiles

Table 6.3: Measurement heights of atmospheric parameters at Lindenberg (grassland) and Hartheim (coniferous forest).

6.2.1 SIMULATIONS FOR THE LINDENBERG GRASSLAND SITE

For Lindenberg, a comparison of soil moisture, soil temperature, and heat fluxes with measurements is possible. The observed soil water content is averaged to three layers from 3–9, 9–27 and 27–81 cm. In the two upper soil layers, the amplitude of soil moisture change due to rainfall is captured well by both models during winter and spring but the absolute values are too low (Fig. 6.2). This is due to the fact that the soil type loamy sand used in the simulations has a lower field capacity than the soil at the measurement site (25 vol.-%). The field capacity of loamy sand in TERRA_LM is 19 vol.-% and in VEG3D 12 vol.-%, which explains the stronger decrease of water content in both simulations at the beginning of the simulation compared to observations.

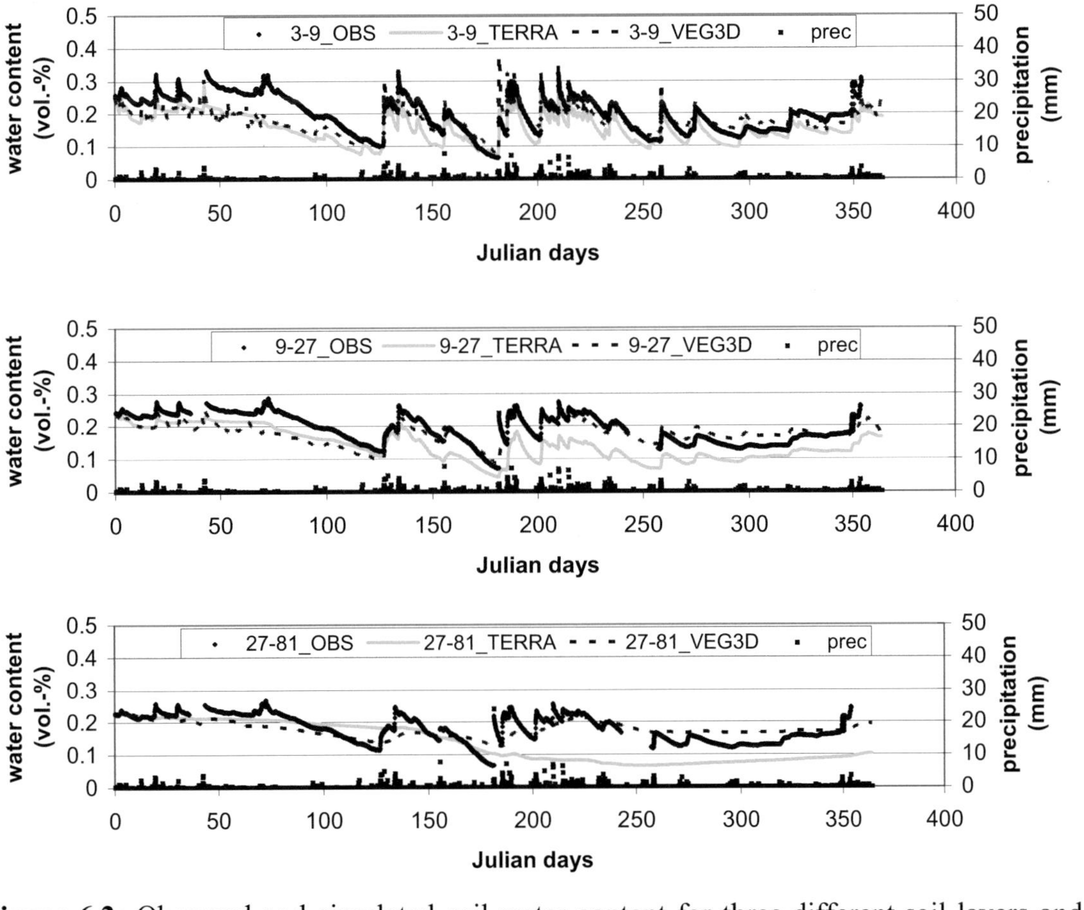

Figure 6.2: Observed and simulated soil water content for three different soil layers and observed precipitation (prec) for the year 2005 at Lindenberg.

During the summer months the water content is simulated reasonably well by both models in the two upper soil layers, with slightly better VEG3D results in the first layer and considerably better VEG3D results in the second layer where maximum differences of up to 10 vol.-% occur in the TERRA_LM simulation compared to the observations (Fig. 6.2). Both models capture the simulation of strong drying out at the beginning of summer and of refilling afterwards. In the lowest soil layer, the models do not simulate the amplitude of the soil moisture change correctly. The amplitude is smoothed by the models and differences up to 15 vol.-% occur between the observations and the TERRA_LM simulation during the summer months. In summer, the VEG3D simulation shows considerable better results than the TERRA_LM simulation for the lowermost layer.

The annual cycle of soil temperature averaged over the depth from 0.07 to 1.42 m is simulated well by VEG3D (Fig. 6.3). The correlation coefficient for the daily mean of soil temperature is 0.998. In the TERRA_LM simulation, the correlation coefficient is 0.992 and there is an underestimation up to 3 K during the summer months. This underestimation is caused by a too strong nighttime cooling (Fig. 6.3). The maximum daytime temperatures are similar for both simulations in the third soil layer (0.07 m) but in the night differences of up to 3 K between the VEG3D and the TERRA_LM simulation occur, which can also be found in the deeper soil layers and cause the too strong cooling compared to observations.

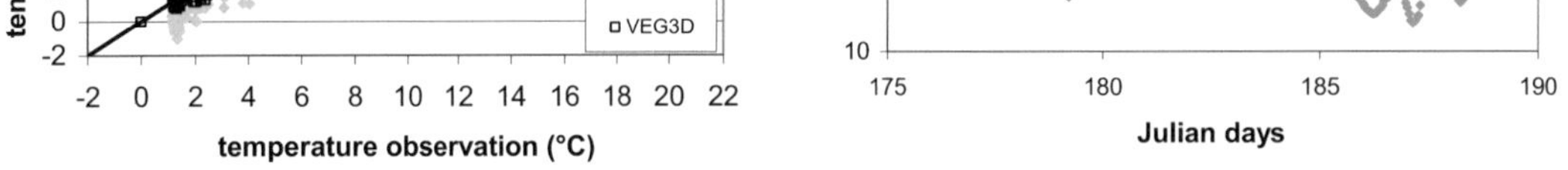

Figure 6.3: Daily mean of observed and simulated soil temperature averaged over the soil layers from 0.07 to 1.42 m for the whole year 2005 (left) and hourly soil temperature of the third soil layer (0.07 m) for the two simulations and the observations for a summer period in 2005 for the Lindenberg site (right).

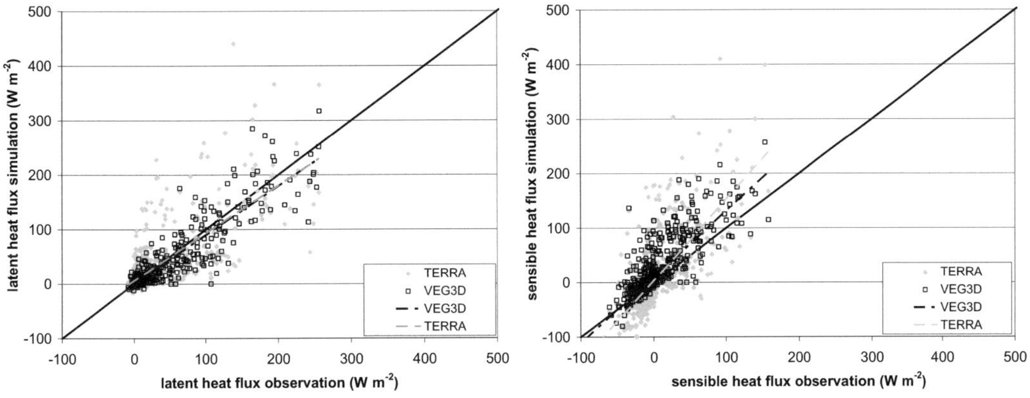

Figure 6.4: Comparison of simulated and observed latent (left) and sensible (right) heat fluxes at Lindenberg site for the year 2005. Points indicate the simulations, dashed lines indicate the linear fit and the solid line is the bisecting line. For purpose of clarity, only every tenth point is shown.

The partition into sensible and latent heat flux is not captured well by both models (Fig. 6.4). The latent heat flux is underestimated by the TERRA_LM simulation and the correlation coefficient is poor (0.55). The VEG3D simulation also underestimates the latent heat flux but the correlation is much better (0.85).

The sensible heat flux is overestimated by both simulations, with higher correlation for the VEG3D simulation (0.65) than for the TERRA_LM simulation (0.50). The latent heat flux is overestimated by the TERRA_LM simulation mostly during the summer months (more than 10 W m^{-2}) where the VEG3D simulation underestimates the latent heat flux (about 5 W m^{-2}) (Fig. 6.5). The sensible heat flux is strongly overestimated by both models (25 W m^{-2}) during the summer months and no model is to be preferred.

During winter, VEG3D results for latent heat flux are quite good but the sensible heat flux is overestimated. The opposite behaviour is observed for the TERRA_LM simulation, which estimates sensible heat flux well but strongly overestimates latent heat flux. The total amount of sensible and latent heat flux for the winter months is captured well in both models, with a maximum difference of 20 W m^{-2} compared to the measurements and a maximum difference of 7 W m^{-2} between the model simulations (Fig. 6.6). In the summer months, the difference between the total amount of heat fluxes in TERRA_LM and VEG3D is around 15 W m^{-2}. This is due to the lower albedo in TERRA_LM (0.16) compared to the albedo of the VEG3D simulation (0.22). The sum of latent and sensible heat flux is overestimated by both models compared to observations (Fig. 6.6) but the overestimation is stronger in the TERRA_LM simulation (30 W m^{-2}) than in the VEG3D run (15 W m^{-2}).

For the chosen land use type and soil type, VEG3D performs better than TERRA_LM with regard to heat fluxes and soil temperatures even for rather low vegetation where one would have expected only small differences between the two models. The main differences in soil water occur in the deeper soil layers, where TERRA_LM underestimates the soil water content considerably.

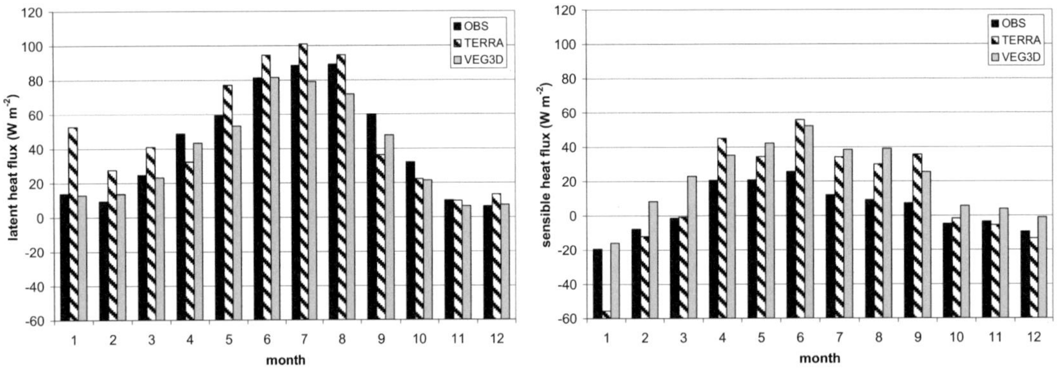

Figure 6.5: Comparison of observed (OBS) and simulated monthly means of latent (left) and sensible heat fluxes (right) at Lindenberg site for the year 2005.

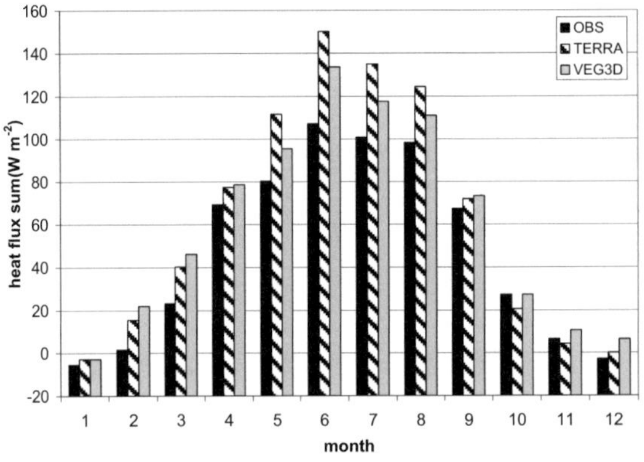

Figure 6.6: Comparison of observed (OBS) and simulated monthly means of the sum of sensible and latent heat fluxes at Lindenberg site for the year 2005.

6.2.2 SIMULATIONS FOR THE HARTHEIM FOREST SITE

For Hartheim, a comparison of soil temperature and energy balance with observations is possible. Unfortunately no water content measurements are available for the year 2001. Unlike the Lindenberg station, the temperature measurements only cover the upper 40 cm of the soil. In this depth range, the average of soil temperature shows good agreement with the observations for the VEG3D simulation except during the summer months, where it overestimates soil temperature by up to 3 K (Fig. 6.7). The TERRA_LM simulation underestimates the temperature during the cold season and overestimates it during the warm season. This overestimation is much higher than the overestimation by the VEG3D simulation. In the upper soil layers the temperature overestimation during daytime reaches 8 K and more in the TERRA_LM simulation compared to observations (Fig. 6.7). In the VEG3D simulation, the additional vegetation layer prevents this heating and the temperature agrees better with the simulations. The maximum difference is only about 5 K. The temperatures during night time are simulated better by the TERRA_LM model. The VEG3D model simulates temperatures that are slightly too high (up to 2 K); the vegetation layer prevents the necessary cooling during nighttime.

In agreement with the observations, the vegetation layer in VEG3D inhibits the strong cooling during winter simulated by TERRA_LM (Fig. 6.8). Unrealistic soil freezing is prevented in the VEG3D simulation as well.

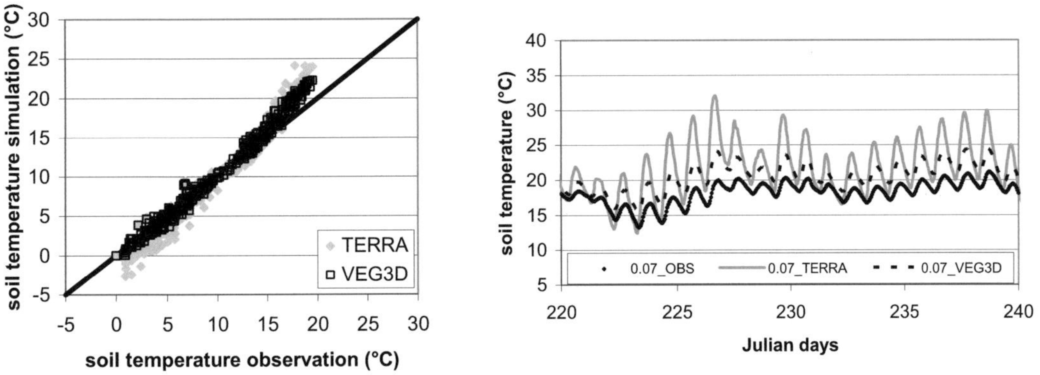

Figure 6.7: Comparison of daily mean simulated soil temperature averaged over the uppermost 40 cm with measurements for the whole year 2001 (left) and comparison of hourly values of observed and simulated soil temperature at 0.07 m for a summer period at the Hartheim site.

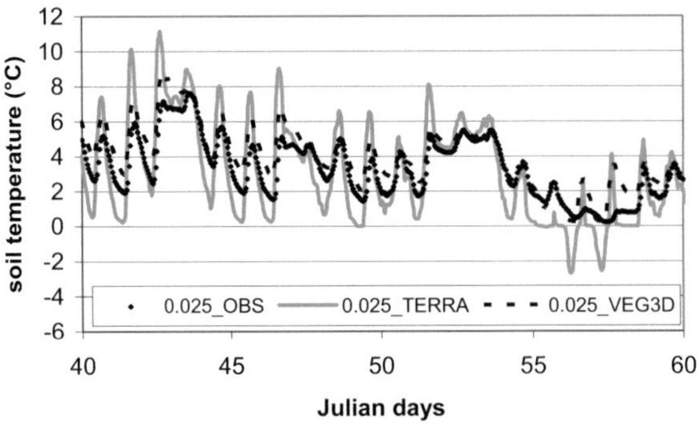

Figure 6.8: Comparison of hourly values of observed and simulated soil temperature at 0.025 m for a winter period in 2001 for the Hartheim site.

The absolute value of thermal net radiation at the surface is higher than in the TERRA_LM simulation (Fig. 6.9). In contrast, the net shortwave radiation is lower than observed in the TERRA_LM simulation (Fig. 6.9). This is caused by an albedo (0.15) higher than the one in the observations (0.11). VEG3D simulates an albedo (0.12) similar to the observed one and consequently the simulation of the net shortwave radiation agrees well with observations. The net thermal radiation is also overestimated by VEG3D but not as strongly as in the TERRA_LM simulation. The correlation of the heat fluxes is worse than at the Lindenberg site. The correlation coefficients for the latent/sensible heat flux are 0.09/0.09 for the TERRA_LM simulation and 0.14/0.13 for the VEG3D simulation. In general, the results from the VEG3D simulation are slightly better than those of the TERRA_LM simulation and both models tend to underestimate sensible and latent heat flux (Fig. 6.10). Both models underestimate the sensible heat flux throughout the whole year, whereas they underestimate the latent heat flux only during winter and spring and show an overestimation of latent heat flux during summer.

The water content cannot be compared to observations but the water up-take by the roots is quite different in both models because of the different root depth. TERRA_LM takes up water mainly from the uppermost six soil layers, whereas VEG3D also takes up water from much deeper layers. Therefore, the water content in the uppermost six layers in the VEG3D simulation does not show as strong a decrease during summer time as it does in the TERRA_LM simulation (Fig. 6.11). The total change of water content over all soil layers is relatively similar until the end of the simulation period.

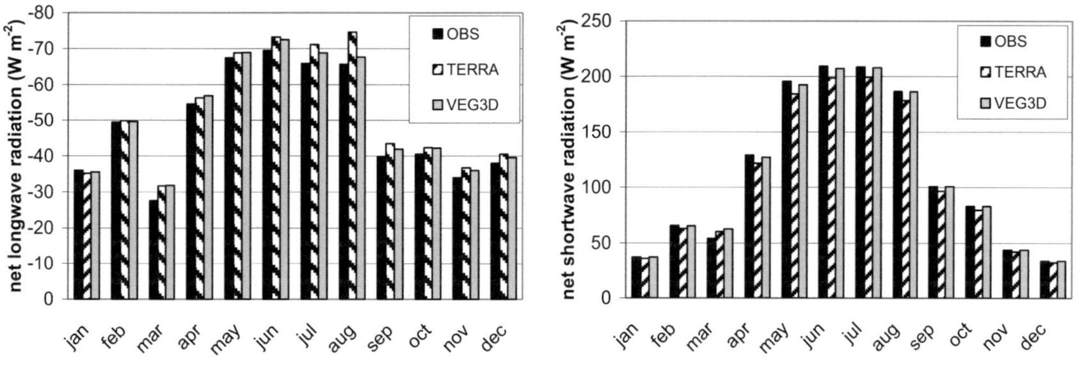

Figure 6.9: Monthly means of surface net longwave (left) and shortwave (right) radiation for the year 2001 at Hartheim site.

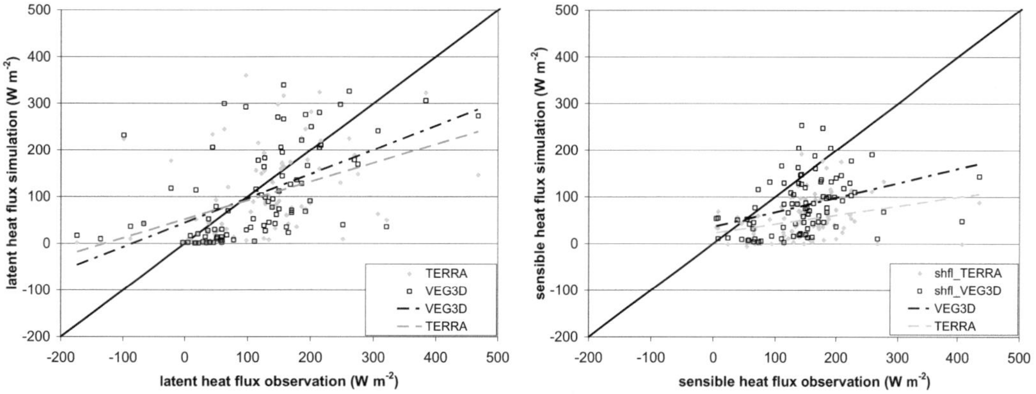

Figure 6.10: Comparison of simulated and observed heat fluxes for the year 2001 at Hartheim site. Points indicate the simulations, dashed lines indicate the linear fit and the solid line is the bisecting line. For clarity, only every tenth point is shown.

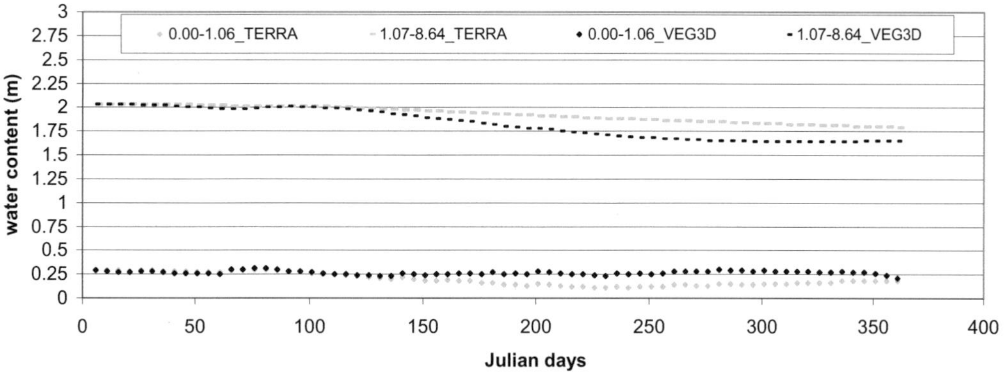

Figure 6.11: Column soil water content for the upper (0.00-1.06 m) and lower (1.07-8.64 m) soil layers in the TERRA_LM and VEG3D simulation for the year 2001 at Hartheim site.

For the Hartheim forest site, VEG3D performs better than TERRA_LM. As expected, the differences between the two simulations, especially in the soil temperatures, are much higher than for the grassland site. Both models show deficiencies in modelling the heat fluxes, due to an incorrect partitioning into latent and sensible heat flux. Apart from uncertainties from applying the Bowen ratio method to forest sites, the parameterisations used in the models may not be suitable for forest canopies (DENMEAD and BRADLEY, 1985) and may thus contribute to the poorer agreement compared to grassland.

6.3 DAILY SIMULATIONS WITH THE ONLINE COUPLED LAND SURFACE SCHEMES

Simulations with the land surface schemes coupled to COSMO-CLM were performed for a day with shallow convection to investigate the influence of the land surface scheme on the boundary layer and the formation of clouds and to check the reliability of the results of the online coupled COSMO-CLM/VEG3D system. GME forecasts were used as driving data.

On 1 June 2002, an intensive observation period of the VERTIKATOR experiment (Vertical Exchange and Orography) took place in Southwest Germany (Barthlott et al., 2006). Radiosondes were launched at two sites in the Rhine Valley and in the Black Forest and a ground-based observation network had been installed to collect information about near-surface meteorological variables such as 2m-temperature and humidity, 10m-wind and heat fluxes. A description of the measurement network and the synoptic conditions on that day can be found in MEISSNER (2004).This data set is used to evaluate the influence of the VEG3D land surface scheme on the planetary boundary layer and on cloud formation. The radio soundings of 6, 12 and 18 UTC are compared with the COSMO-CLM simulations coupled online with TERRA_LM and VEG3D for the stations Freistett and Horb (stations are depicted in Fig. 9.11).

At Freistett, the vertical profiles between the two simulations vary even at 6 UTC (Fig. 6.12). The VEG3D simulation shows lower specific humidity and higher temperature than the TERRA simulation up to 950 hPa. Neither of the two models simulates the strong inversion in temperature in the lowest 50 hPa, so no model gives better results than the other. For humidity, the TERRA simulation shows better results than the VEG3D simulation at 6 UTC. At 12 UTC both simulated temperature profiles are quite similar and underestimate the observed profile by up to 1 K. The first inversion is not simulated in both runs. For specific humidity, both profiles match well up to a height of 950 hPa. Above this height, humidity is reduced in the VEG3D simulation up to a height of 900 hPa and increased above 900hPa, compared to the TERRA simulation. The absolute values of specific humidity agree better with observed values for the VEG3D simulation but the boundary layer height is lower than in the TERRA simulation, which is not in accordance with the observations.

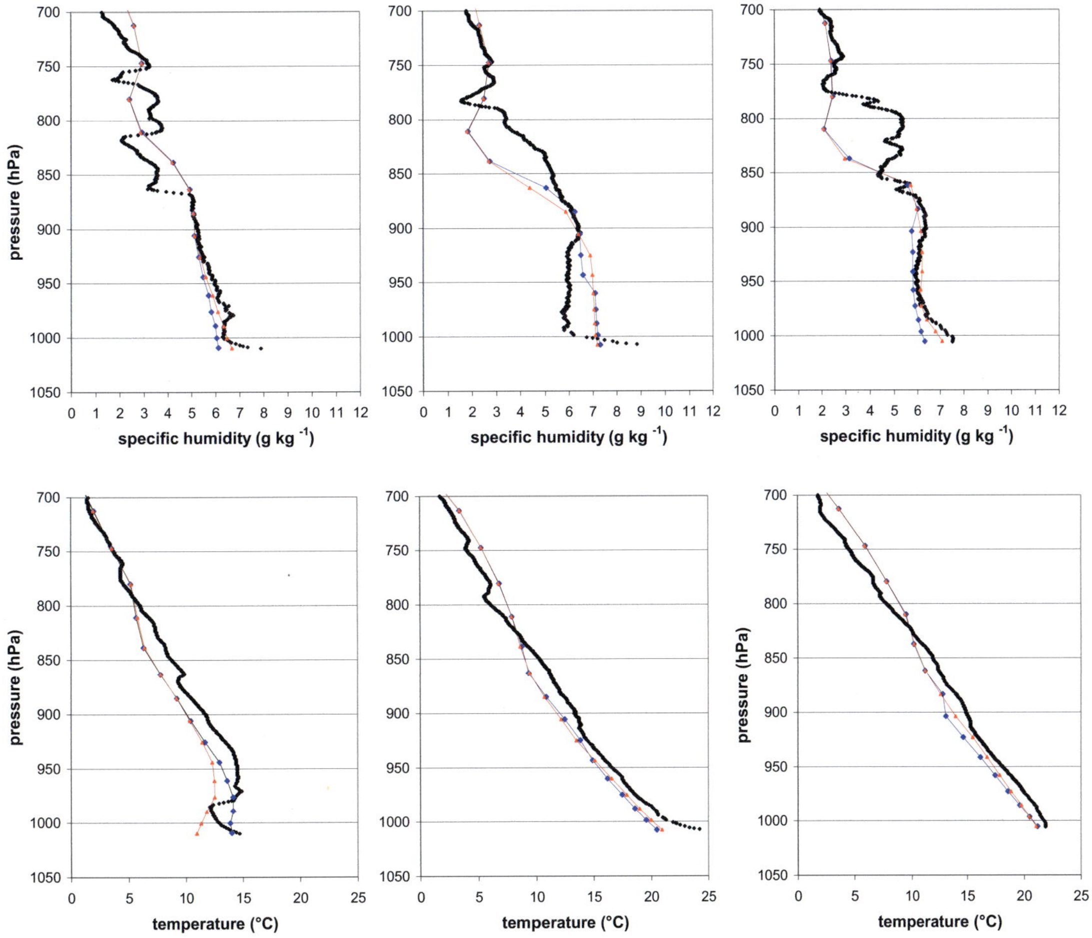

Figure 6.12: Vertical profiles of specific humidity and temperature of observations (black dots), TERRA simulation (red line) and VEG3D simulation (blue line) at the Freistett site for 1 June 2002 at 6 (left), 12 (middle) and 18 UTC (right).

At 18 UTC, the temperature profile in the VEG3D simulation is slightly colder than the TERRA profile, which disagrees with the observation. However, the VEG3D simulation is able to simulate the inversion, which is not simulated in the TERRA run. Both profiles agree well with the observations for specific humidity, with slightly higher humidity values and therefore better agreement with the observations in the TERRA simulation.

At Horb, the differences at 6 UTC are not as pronounced between the two simulations as in Freistett (Fig. 6.13). Both specific humidity profiles are rather similar. The temperature is simulated better in the VEG3D run but the inversion in the lowermost 50 hPa is missing in both simulations.

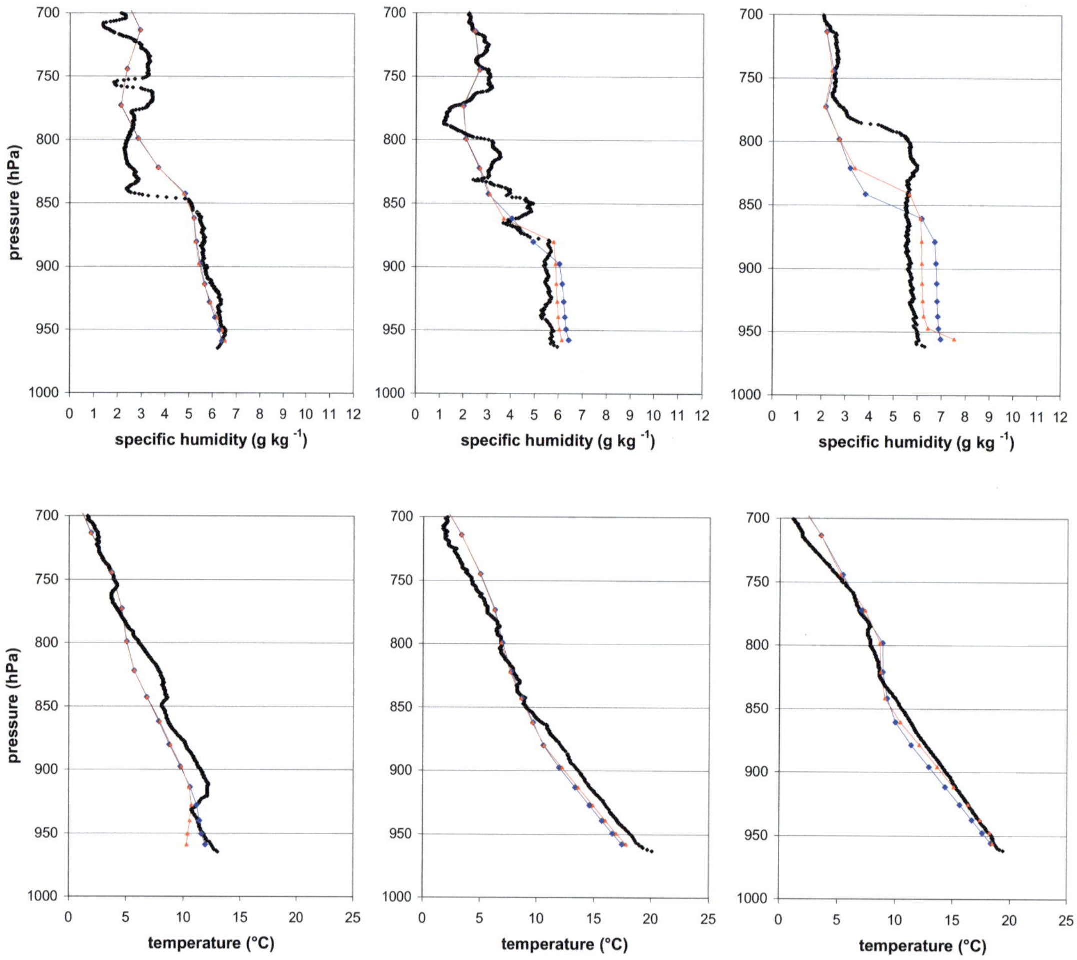

Figure 6.13: Vertical profiles of specific humidity (top) and temperature (bottom) of observations (black dots), TERRA simulation (red line) and VEG3D simulation (blue line) at the Horb site for 1 June 2002 at 6 (left), 12 (middle) and 18 UTC (right).

At 12 UTC, the temperature profile between both simulations is quite similar, as was already found for Freistett. The humidity in the boundary layer shows higher values in the VEG3D simulation than in the TERRA simulation. Compared to the observations, both models simulate a humidity amount that is too high in the boundary layer and too low above the boundary layer. As in Freistett, the boundary layer height is given more realistically by the TERRA simulation. These differences in vertical humidity profile are enhanced until 18 UTC. In the VEG3D profile, the surplus of specific humidity compared to observations is located nearer to the surface than in the TERRA simulation and leads to an overestimation of specific humidity by up to 1.5 g kg^{-1} compared to the observations. In the TERRA simulation, the humidity is transported in higher altitudes and therefore the values of humidity agree better with the observations up to 870 hPa.

The cloud development is quite similar in both simulations (Fig. 6.14). As can be seen from the satellite images, the cloud development at 12 UTC agrees well with the observed one. As observed, clouds form over the northern and the southern part of the Black Forest in both simulations. The intensity of cloudiness is similar in both simulations. At 16 UTC, strong cloud cover is simulated only for the southern Black Forest by both models and the cloudiness is slightly higher in the VEG3D simulation. In the Northern Black Forest the cloud-covered area is even reduced in the simulations. This disagrees with the observations, where the cloud cover in the Northern Black Forest increases.

The heat fluxes show differences, especially in the Rhine Valley and the Vosges Mountains (Fig. 6.15). The latent heat flux values in the Rhine Valley are quite low (<100 Wm^{-2}) in the TERRA simulation and quite high (up to 500 W m^{-2}) in the Vosges Mountains. In reverse, the sensible heat flux is too low over the Vosges Mountains (<100 W m^{-2}) and too high in the Rhine Valley (400 W m^{-2}). For forested areas, such as the Vosges Mountains, one would expect a bowen ratio (ratio of sensible to latent heat flux) of 1.0 and for areas with grassland, as the Rhine Valley, one would expect a bowen ratio of 0.7 (ROST and MAYER, 2006). In the TERRA simulations we get values of 0.25 for forest and 4 for grassland. In the VEG3D simulation, the difference between the heat fluxes for forested and cultivated areas is smaller but clearly visible. The amount of 300 W m^{-2} and 200 W m^{-2} for latent heat flux over forest and cultivated areas and 200 W m^{-2} and 100 W m^{-2} for sensible heat flux seems adequate compared to literature values (ROST and MAYER, 2006).

The variability in the radiation balance is much higher in the VEG3D simulation (Fig. 6.16). The values of the radiation balance are higher over the forested areas and lower over the cultivated areas due to the lower albedo over forest, and maximum differences of 120 W m^{-2} are observed. In the TERRA simulation, this difference is not as pronounced and maximum differences of 80 W m^{-2} are observed.

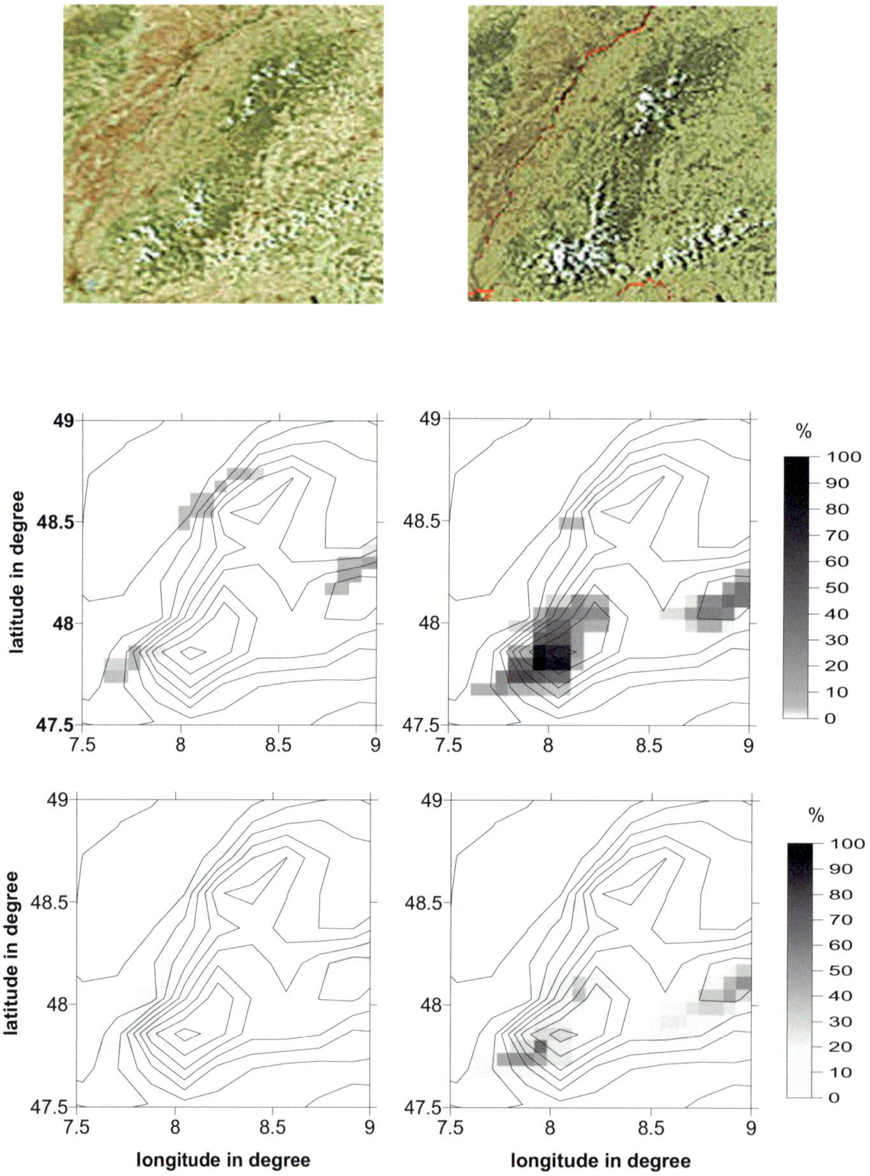

Figure 6.14: Satellite images of cloud cover (top) and simulated cloud cover in percent of grid box from VEG3D simulation (middle) and TERRA simulation (bottom) for 12 UTC (left) and 16 UTC (right) at 1 June 2001.

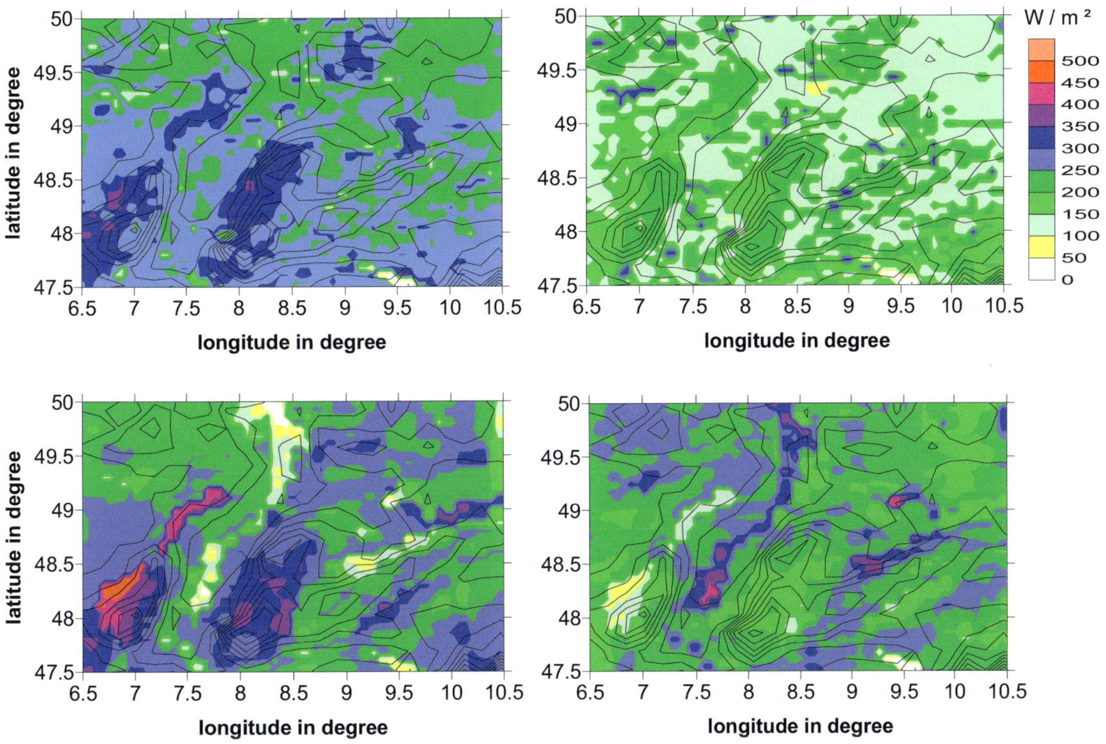

Figure 6.15: Latent (left) and sensible (right) heat flux for 1 June 2002 at 12 UTC for the VEG3D simulation (top) and the TERRA simulation (bottom).

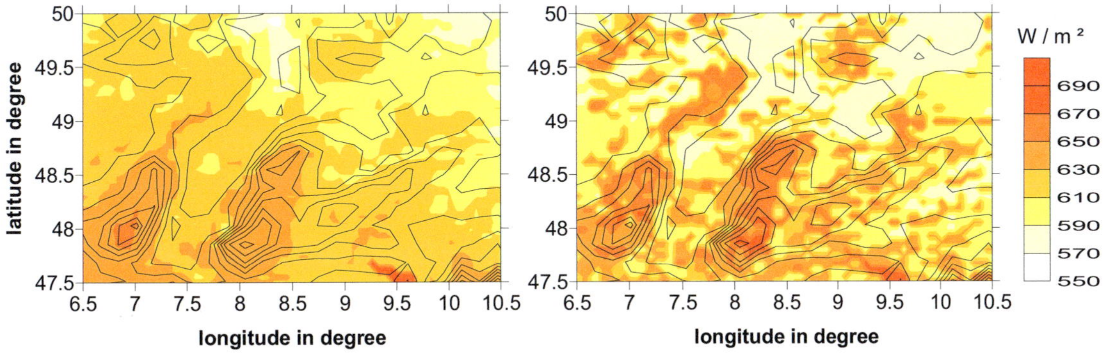

Figure 6.16: Radiation balance for 1 June 2002 at 12 UTC for the TERRA simulation (left) and the VEG3D simulation (right).

For a more detailed comparison, the results of the two simulations were evaluated by observed station data. For this comparison, the observations at the Sasbach and Musbach sites were used. These were the only two sites measuring the radiation balance as well as the heat fluxes and the 2m-temperature.

At Sasbach site, the values of sensible heat flux are simulated well by the VEG3D run but they are about 100 W m^{-2} too high in the TERRA run (Fig. 6.17). The values of latent heat flux between both models agree well and an earlier decrease in the diurnal cycle can be observed in the VEG3D simulation. Unfortunately, no latent heat flux was measured on that day. The radiation balance is similar for both simulations and the daily variation agrees well with the observations (Fig. 6.17). During night time, both models underestimate the radiation balance by about 50 W m^{-2}. At noon, an overestimation of about 50 W m^{-2} is found for the TERRA run and about 10 W m^{-2} for the TERRA run. The diurnal cycle of 2m-temperature is well represented by both models, with slightly higher temperatures than observed during the night (Fig. 6.17). The amplitude is slightly too small in both simulations. This leads to an underestimation of temperature at noon in the VEG3D simulation and to an overestimation in the TERRA simulation. The values of near-surface specific humidity are similar during nighttime but during the day the value is about 2 g kg^{-1} higher in the VEG3D simulation (Fig. 6.17).

At the Musbach site, the simulated heat fluxes are similar in both models with slightly higher values for latent heat flux in the TERRA simulation (30 W m^{-2}) and slightly higher values of sensible heat flux for the VEG3D simulation (20 W m^{-2}) (Fig. 6.18). The modelled values for latent and sensible heat flux are higher than observed by about 50 W m^{-2}. The radiation balance is also higher than the observed one by about 70 W m^{-2} (Fig. 6.18). Adding up radiation balance and heat fluxes a maximum soil heat flux of 120 W m^{-2} is obtained for both model runs, which is similar to the observed one (100 W m^{-2}). The amplitude of the diurnal cycle of 2m-temperature is reproduced well by both models. For the TERRA simulation, realistic results are obtained compared to the observations but for the VEG3D simulations the values are about 2 K too low (Fig. 6.18). This is due to the land use class forest, which is used at this grid point in the model. A detailed explanation for this underestimation is given in subsection 6.4.2. The diurnal cycle of near-surface specific humidity is quite similar in both models and differences are lower than 1 g kg^{-1}.

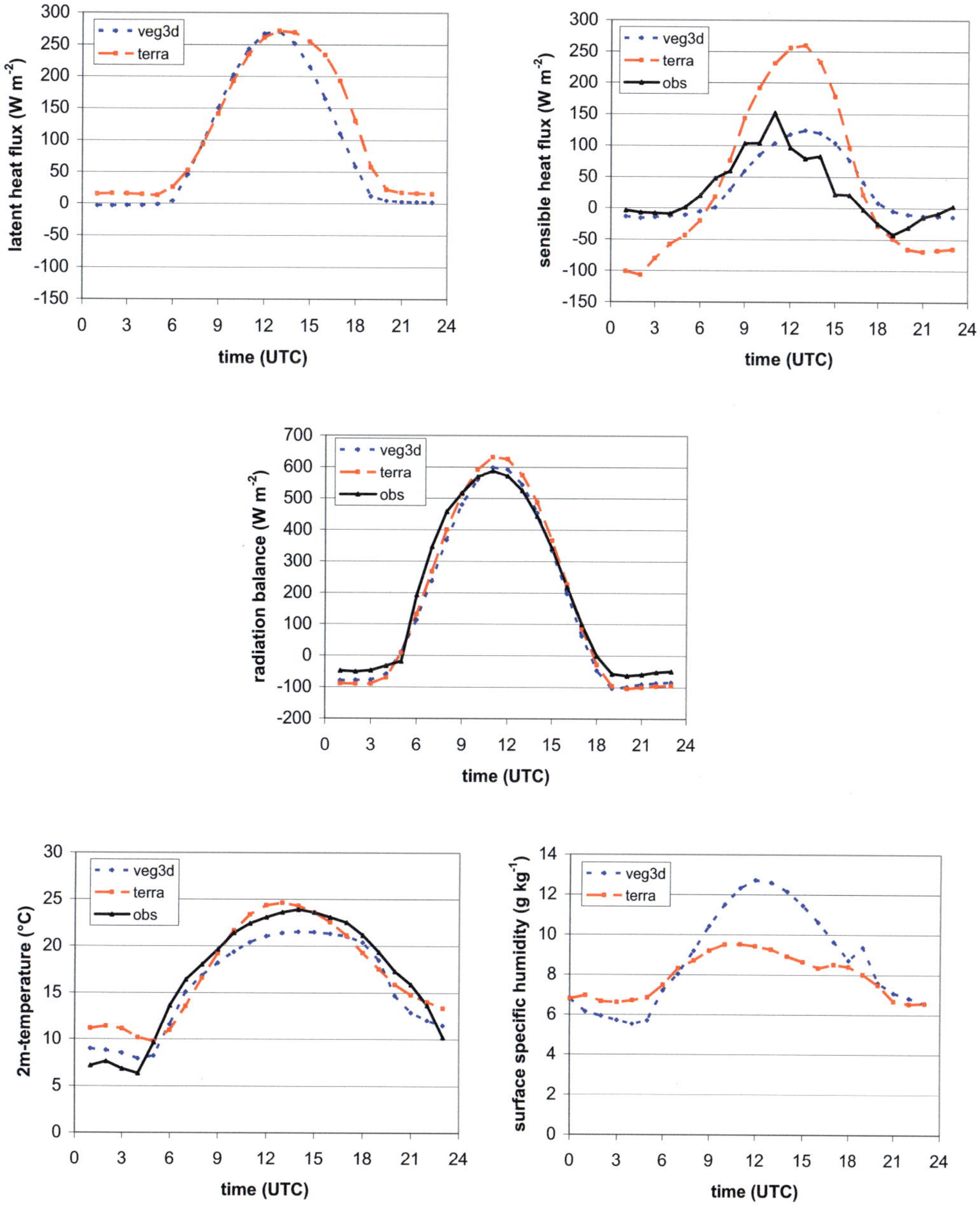

Figure 6.17: Latent and sensible heat flux (top), radiation balance (middle), 2m-temperature and near-surface humidity (bottom) for observations (obs) and VEG3D and TERRA simulation for 1 June 2002 at the Sasbach site.

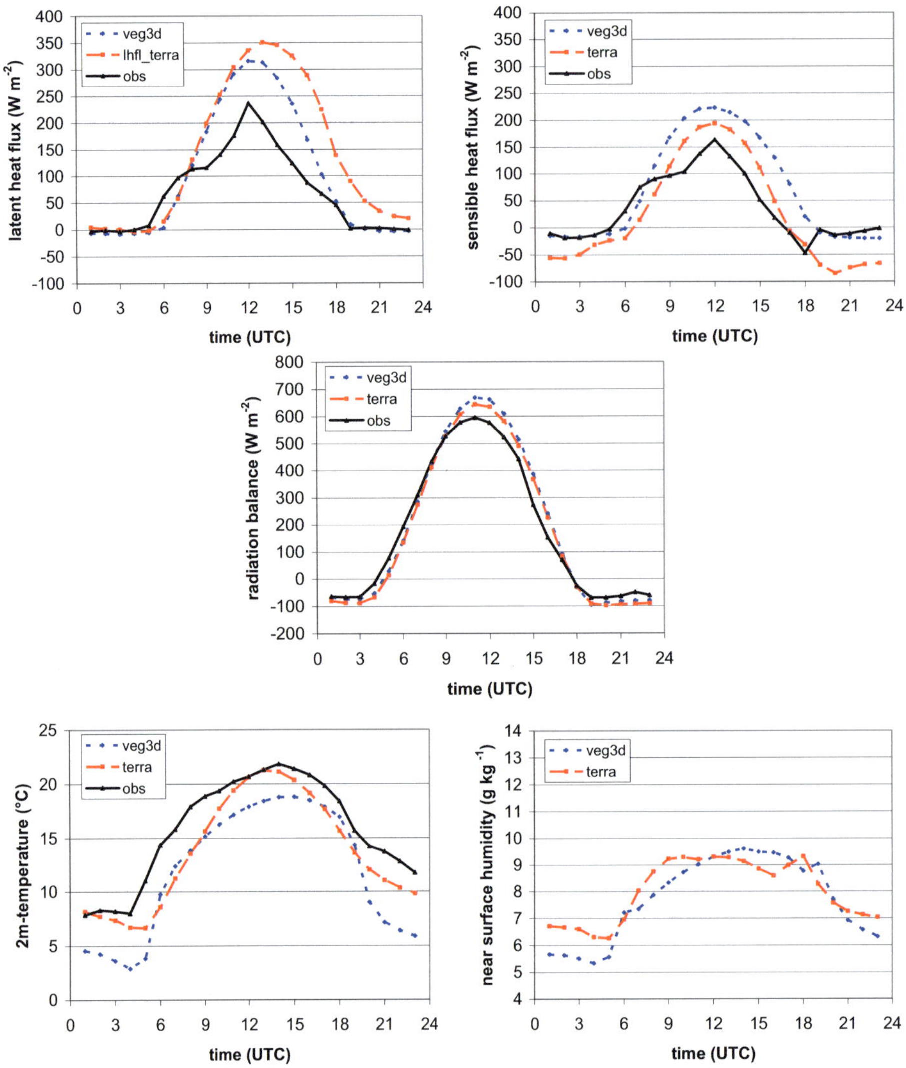

Figure 6.18: Latent and sensible heat flux (top), radiation balance (middle), 2m-temperature and near-surface humidity (bottom) for observations (obs) and VEG3D and TERRA simulation for 1 June 2002 at the Musbach site.

Both models are able to reproduce the vertical profiles and the diurnal cycle of heat fluxes, radiation balance and 2m-temperature but differences between the two simulations are clearly visible. For a first comparison and for a check of qualitative reliability daily simulations are adequate and both models have proven this reliability. Performing daily simulations is the method normally used to test new parameterisations. Due to the necessary spin-up time of the soil model, which normally takes at least several months, the evaluation of longer time scales is essential for an adequate comparison between the two online coupled land surface schemes.

6.4 Simulations with the two land surface schemes coupled online with COSMO-CLM for the year 2001

Simulations with the two SVATs coupled to COSMO-CLM were performed for the year 2001 using GME analysis data (MAJEWSKI et al., 2002) as driving data. From the water balance point of view, 2001 can be considered an average year, without extended periods of draught or wetting. The advantage of GME data compared to ERA or NCEP reanalysis data is its higher resolution (about 60 km). Therefore, the COSMO-CLM simulations with 7 km grid size can be driven directly without a nesting step. This corresponds with the way the operational COSMO-EU weather forecast model is run. The simulations were run from the first of December 2000 to January 2002. Due to the proximity to the Alps, the area of simulation is much larger than the investigation area and includes the whole alpine region (see subsection 4.3.1) (Fig. 6.19). Model results are evaluated for the Southwest of Germany (Fig. 6.19). The simulations are compared with observational data from 55 weather stations operated by the German Weather Service (DWD) that were pre-processed by the PIK Potsdam (ÖSTERLE et al., 2006).

COSMO-CLM version 4.0 is used, with a horizontal resolution of 0.0625° and a time step of 40 seconds. The parameter settings and parameterisations are the same as they are for the ERA/7km run described in section 4.1. For the simulation with TERRA_LM only the lowest soil layer in TERRA_LM is a hydrological passive layer. All other layers are hydrological active layers.

In addition to the quantities considered in section 6.2, annual precipitation and near-surface temperatures are also compared. Annual precipitation, evaporation sums, soil water content and soil temperature are compared for the entire investigation area. Maximum, minimum and mean 2m-temperature and precipitation are compared at observation sites.

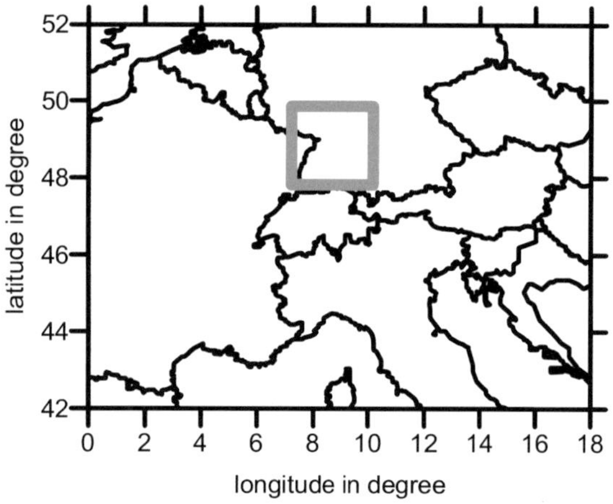

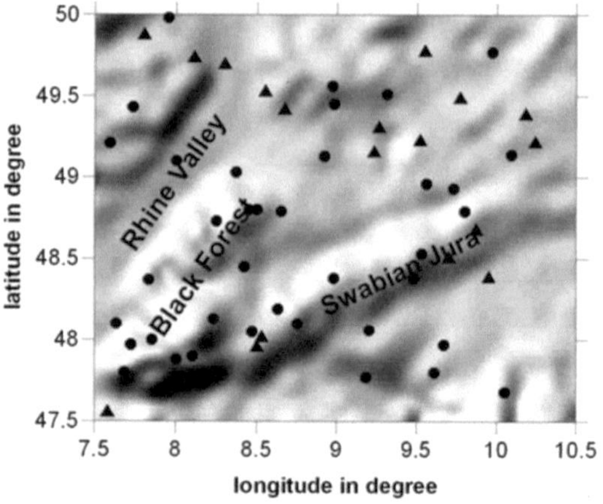

Figure 6.19: Left: Simulation domain with investigation area (grey rectangle). Right: Investigation area with measurement stations. Forest stations are indicated by black dots and non-forest stations by black triangles (for explanation see text sec. 6.4.2)

6.4.1 Comparisons for the whole investigation area

The observed annual precipitation sum in the investigation area ranges from 500 mm in the Rhine Valley to 2000 mm in the Black Forest. A lee effect with a pronounced reduction of the precipitation sum can be observed in the east of the Black Forest. For the Swabian Jura precipitation is about 900 – 1400 mm: further north it is about 600 – 1000 mm in the lowlands and 1000 – 1400 mm at elevated sites. Both simulations reproduce this precipitation distribution qualitatively but the total amounts at the elevated sites are overestimated by up to 600 mm. A comparison of the annual precipitation sum between the two simulations is shown in Fig. 6.20. The total precipitation amount in the VEG3D simulation is higher at the western slopes of the Black Forest (which is the windward side for the prevailing westerly flows) and also over the whole area of the Swabian Jura (Fig. 6.20). In the area of the Black Forest and the region north of it, the precipitation amount is generally lower than in the TERRA_LM simulation and differences reach values of up to 150 mm. These patterns can also be found in the annual grid scale precipitation (Fig. 6.20). For the convective (subgrid) precipitation, patterns are similar to the total precipitation in the southern part of the investigation area; for the northern part, higher convective precipitation in VEG3D is offset by grid scale precipitation resulting in a total precipitation lower than in the

TERRA_LM simulation for that region (Fig. 6.20). The area average of total precipitation is about 50 mm higher in the TERRA_LM simulation (Tab. 6.4) due to higher grid scale precipitation (55 mm). The convective precipitation amount is slightly higher in the VEG3D simulation (5 mm). The monthly grid scale precipitation is higher in the TERRA_LM simulation all year, whereas the monthly convective precipitation is only from May to July. The rest of the year, monthly convective precipitation amount is equal to or lower than it is in the VEG3D simulation.

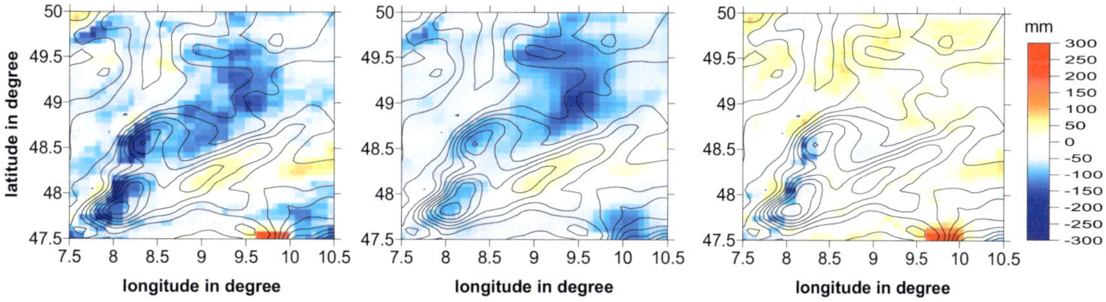

Figure 6.20: Difference in annual precipitation sum for 2001 between VEG3D and TERRA_LM simulation for total precipitation (left), grid scale precipitation (middle) and convective precipitation (right).

Precipitation (mm)	TERRA_LM	VEG3D
Total	1271	1222
Grid scale	888	834
Convective	382	387

Table 6.4: Area average of total, grid scale and convective precipitation for TERRA_LM and VEG3D simuation for the year 2001.

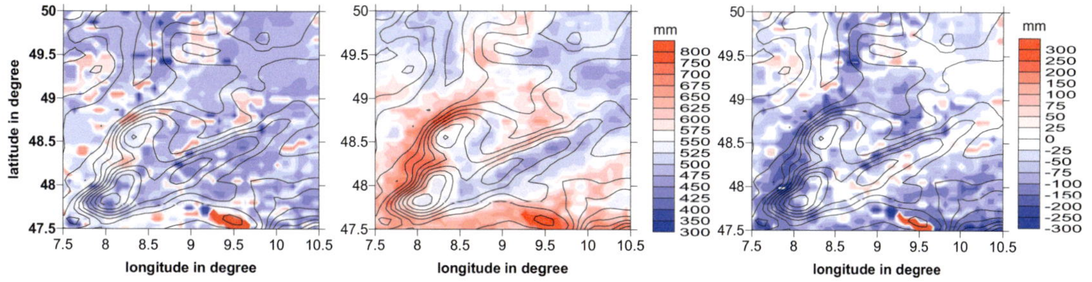

Figure 6.21: Annual evapotranspiration sum for VEG3D simulation (left) and TERRA_LM simulation (middle) and difference between VEG3D and TERRA_LM annual evapotranspiration sum (right) for the year 2001.

The patterns of the annual evapotranspiration sum are similar between the two simulations in most regions (Fig. 6.21). Large differences of up to 160 mm per year between the two simulations can be found for transition areas where different land use classes are mixed within one grid box in the TERRA_LM simulation and there is mainly a mixture of forest, fields and grassland (Fig. 6.21). Evapotranspiration rates in the TERRA_LM simulation are too high in these areas compared to climatological averages given in the Hydrological Atlas Deutschland (HAD, 2003), the overestimation being in the range of 50 mm. On the other hand, the values obtained by VEG3D are slightly too low in these areas (underestimation of about 25 mm). For the rest of the investigation area the amount of evapotranspiration is similar between the two simulations and the values in the Hydrological Atlas. However, since there is no observed evapotranspiration data for 2001, apart from assuming that 2001 is close to the climatological average, a quantitative assessment of the quality of the simulations is not possible. To compare the behaviour of the soil temperature, the soil temperatures are averaged over the uppermost nine soil layers (a depth of 8.60 m) and over the whole investigation area. The annual amplitude is similar in both simulations (Fig. 6.22). The VEG3D soil temperatures are higher by up to 1 K throughout the summer period. During winter, the TERRA_LM simulation is sometimes warmer than the VEG3D simulation for some periods over several days due to differences in snow cover. For the soil water content the amount over the uppermost nine layers, which are the active soil layers in TERRA_LM, is summed up and is then averaged over the whole investigation area. The total soil water content decreases strongly during the first month in the TERRA_LM simulation (Fig. 6.22). This is due to the runoff formulation discussed in section 6.1. A second strong decrease can be found during the summer months where the total water content is reduced by about 0.3 m. This is mainly due to evapotranspiration. The decrease in summer due to evaporation is lower in the VEG3D simulation by about 0.2 m. The decrease of water content at the beginning of the simulation is not as strongly visible in the VEG3D simulation as it is in the TERRA_LM simulation due to the different runoff formulation. The difference between the two simulations is about 0.2 m in total soil water content at the beginning of the vegetation period; at the end of the simulation period the difference in total soil water content is about 0.35 m for the uppermost nine soil layers.

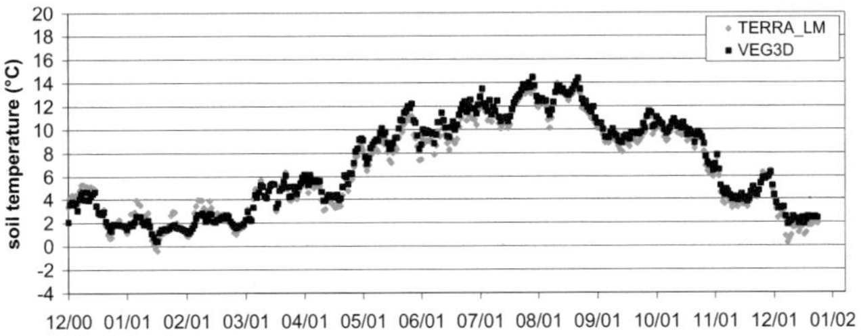

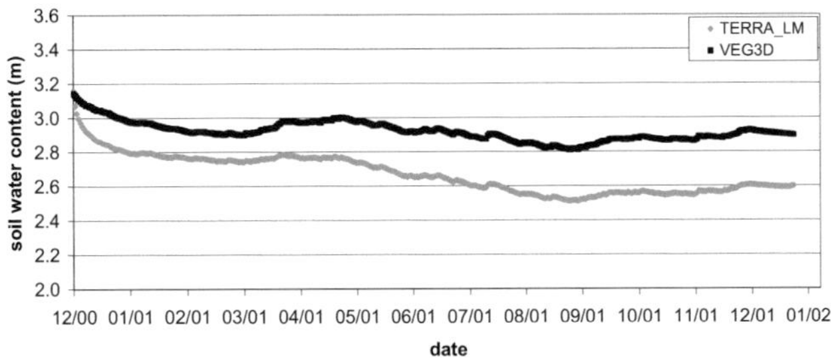

Figure 6.22: Annual cycle for the mean soil temperature (top) and the mean total water content of the first nine soil layers (bottom) averaged over the investigation area.

Table 6.5 shows the components of the soil water balance from the start of the simulation in December 2000 until end of December 2001, averaged over all land points within the investigation area. It is noticeable that the evaporation and the precipitation sum of the TERRA_LM simulation is about 61 and 56 mm per grid box higher than it is in the VEG3D simulation. The total runoff in the TERRA_LM simulation is about 4 mm higher and surface runoff about 21 mm lower compared to the VEG3D results, whereas ground runoff is higher by about 25 mm. The modelled change in soil water content is larger in VEG3D in contrast to the findings in Fig. 6.22. The difference in soil water content visible in the nine active soil layers is enhanced by the soil water decrease in the tenth layer of VEG3D. In general, the water balance of both models is similar, especially for runoff.

Model run	Evaporation (sum)	Surface runoff (sum)	Ground runoff (sum)	Precipitation (sum)	Water content (difference)
TERRA_LM	577	228	1077	1341	-563
VEG3D	516	249	1052	1285	-610

Table 6.5: Water balance in mm averaged over all land points within the investigation area for the period from December 2000 to December 2001.

6.4.2 Comparisons with data from observation sites

For climate simulations, the commonly compared meteorological variables are near-surface temperature and precipitation. To investigate the influence of the land surface scheme on these parameters, the annual minimum, maximum, mean 2m-temperature and the precipitation amount of the two simulations are compared with observations from 55 weather stations. The observed station data is compared to a weighted mean of the values for the surrounding four grid boxes of the simulations.

The ability of VEG3D to simulate temperature in line with observations differs strongly between stations with forest land use and stations with other land use. This is due to the fact that weather stations are often located on grassland sites near a forest and not within the forest. Therefore, the land use near the weather station is often not forest, even if the majority of the land use within the corresponding grid box is forest and is used by the model. It is obvious that such a comparison will show large differences. Therefore, the comparison is partitioned into two parts: in the first section, sites where the surrounding four grid boxes are not classified as forest are compared. This restricts the number of stations for the comparison of 2m-temperatures to 15 stations (see Fig. 6.19). In the second category, the 40 forest stations are compared with the simulations. The same partitioning is used to evaluate the precipitation amount.

The comparison of the maximum and mean 2m-temperatures and the precipitation amount shows similar results for the TERRA_LM and VEG3D simulation for the non-forest sites (Fig. 6.23). The maximum differences between the two simulations are about 0.3 K and 150 mm. The correlation of the annual mean and maximum 2m-temperature and the annual precipitation amount with the observations are slightly better for the VEG3D simulation. The non-forest sites are mainly situated in areas where the total precipitation amount decreases in the VEG3D simulation compared to the TERRA_LM simulation (Fig. 6.19, Fig. 6.20). The observed total precipitation amount is also lower in these areas explaining the better results obtained with the VEG3D, which are expressed by a lower intercept and a slope closer to one (Tab. 6.6). Larger differences between the two simulations can be found for the minimum 2m-temperature (Fig. 6.23). Temperatures in the VEG3D simulation are up to 1 K lower compared to the TERRA_LM simulations. The TERRA_LM simulation generally overestimates the minimum temperature so that the results with coupled VEG3D model agree better with observations than the results with the coupled TERRA_LM model and the intercept is about 1 K lower for the VEG3D simulation (Tab. 6.6).

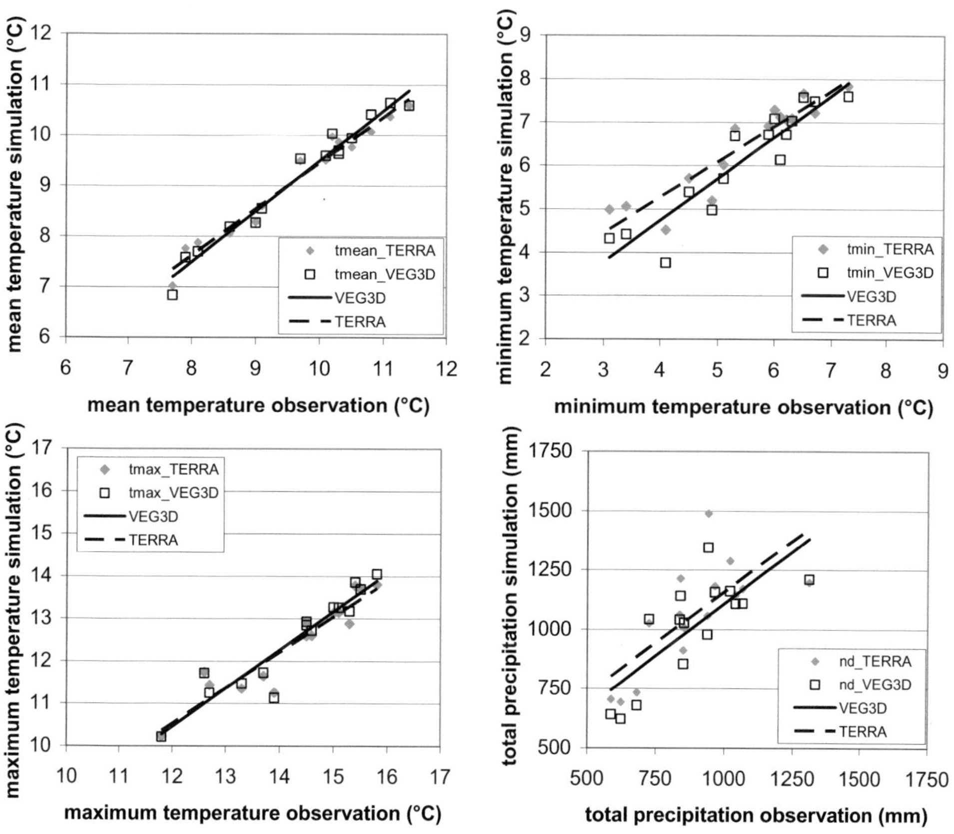

Figure 6.23: Observed versus simulated annual mean, minimum and maximum 2m-temperature and annual precipitation amount for 15 non-forest sites for the year 2001. Lines indicate the least squares fit.

	a_VEG3D	a_TERRA	b_VEG3D	b_TERRA	r_VEG3D	r_TERRA
Mean temperature	0.997	0.912	-0.478	0.333	0.970	0.967
Minimum temperature	0.958	0.818	0.907	1.998	0.856	0.859
Maximum temperature	0.901	0.836	-0.348	0.508	0.891	0.875
Precipitation	0.879	0.860	229.150	297.750	0.590	0.524

Table 6.6: Linear correlation coefficient (r), slope (a), and intercept (b) of the least squares fit between observed and simulated meteorological variables for 15 non-forest sites.

For forest sites, the annual mean 2m-temperature is strongly underestimated by the VEG3D simulation, which is not as pronounced in the TERRA_LM simulation (Fig. 6.24). This underestimation is due to temperatures in the summer months that are too low. During summer, the canopy prevents the near-surface air from heating. Therefore, an underestimation compared to the TERRA_LM simulation and the observations occurs.

The precipitation amount is overestimated for most of the forest stations by both simulations but the differences between the two simulations are small (< 50 mm). A reduction of precipitation amount and therefore better results compared to the observations at these stations can be observed in the VEG3D simulation. The largest differences between the two model results are found for stations where the modelled precipitation amount is lower than the observed one. Here differences up to 100 mm occur and the underestimation is stronger by VEG3D. The majority of these stations are located in the eastern part of the Black Forest where precipitation is reduced compared to the TERRA_LM simulation, as already discussed for Fig. 6.20. The linear regression statistics given in Tab. 6.7 show clearly that neither models perform as well for forest as for non-forest observation sites and that the agreement of the VEG3D simulations with observations is less than for the TERRA_LM simulation, for the reasons discussed above.

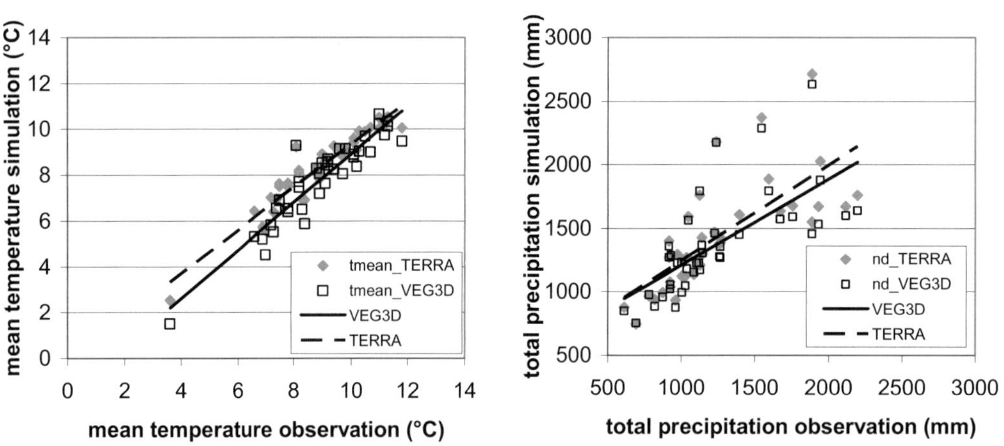

Figure 6.24: Observed versus simulated annual mean 2m-temperature (left) and annual precipitation (right) for 40 forest sites. Lines indicate the least squares fit.

	a_VEG3D	a_TERRA	b_VEG3D	b_TERRA	r_VEG3D	r_TERRA
Mean temperature	1.05	0.94	-1.56	-0.05	0.86	0.90
Precipitation	0.68	0.75	530.14	494.69	0.48	0.54

Table 6.7: Linear correlation coefficient (r), slope(a), and intercept (b) of the least squares fit between observed and simulated meteorological variables for 40 forest sites.

6.5 LONG-TERM EVALUATION

To determine whether the findings for the comparison of online-coupled VEG3D and TERRA_LM simulations obtained for the year 2001 are also valid for longer periods and different driving data, a COSMO-CLM simulation coupled with VEG3D was performed for the period 1988-2001. For this simulation, the model setup described in section 4.1 was used, with 7 km horizontal resolution, ERA-40 driving data for the 50 km run and a large domain size. The only difference from the ERA-40/7 km simulation, which was evaluated in subsection 4.3.2, is the use of the VEG3D land surface scheme instead of the TERRA_LM scheme. The two simulations are compared to determine the influence of the land surface scheme for longer timescales.

In the first subsection, the simulation with VEG3D for the period from 1991 to 1995 is evaluated, as it was done for the changed convection parameterisations and the changed time level scheme in subsection 4.3.3. The variability in simulation results induced by a different land surface scheme is compared to the variability obtained by using other physical and dynamical parameterisations. In the second subsection, the period from 1991–2000 is evaluated as it was done for the year 2001. In the following the ERA-40/7km run with TERRA_LM land surface scheme is referred to as TERRA run and the simulation with VEG3D land surface scheme as VEG3D run.

6.5.1 Evaluation for the period from 1991 to 1995

As shown in subsection 4.3.3, changes in soil moisture content, active soil layers and greenhouse gas amount only have a small effect on the simulation results. Therefore, the simulation results obtained by using VEG3D are compared to the simulations with the Kain-Fritsch scheme (kf), the Runge-Kutta scheme (rk), the TERRA run (ref), and the observations (mes). For 2m-temperature the comparison is made for the average of the 15 non-forest stations. For precipitation the comparison is made for the average of all 55 stations within the investigation area.

The differences between the averaged annual mean 2m-temperature of the TERRA run and the VEG3D run are similar to the differences between the other two schemes and the TERRA run (about 0.2 K) (Fig. 6.25). In four of the five years the temperature values of the VEG3D simulation show the same tendencies as the results obtained with the Kain-Fritsch scheme and underestimate the temperature values compared to the reference run. In these years, the reference run agrees better with the observations than the VEG3D simulation.

Smaller averaged annual precipitation amounts than in the TERRA run can be observed for the VEG3D run in four of five years (not corresponding to the four years where temperature is underestimated). In these years the VEG3D run fits better with the observations than the TERRA run (Fig. 6.25). The variance in precipitation amount (up to 60 mm) produced by the VEG3D simulation is smaller than the variance produced by the other two simulations with changed parameterisations (up to 200 mm).

Comparing the station average of monthly mean 2m-temperature averaged over the period from 1991 to 1995, the VEG3D produces higher temperatures and therefore better simulation results compared to observations than the TERRA run for nearly all months, except the summer months from July to October (Fig. 6.26). Here differences up to -0.4 K occur compared to the TERRA run. The monthly average differences between the VEG3D run and the TERRA run for precipitation are much smaller than the ones between the Kain-Fritsch scheme and the Runge-Kutta scheme (Fig. 6.27).

The highest differences from the TERRA run are produced by the VEG3D scheme for temperature and by the Runge-Kutta scheme for precipitation.

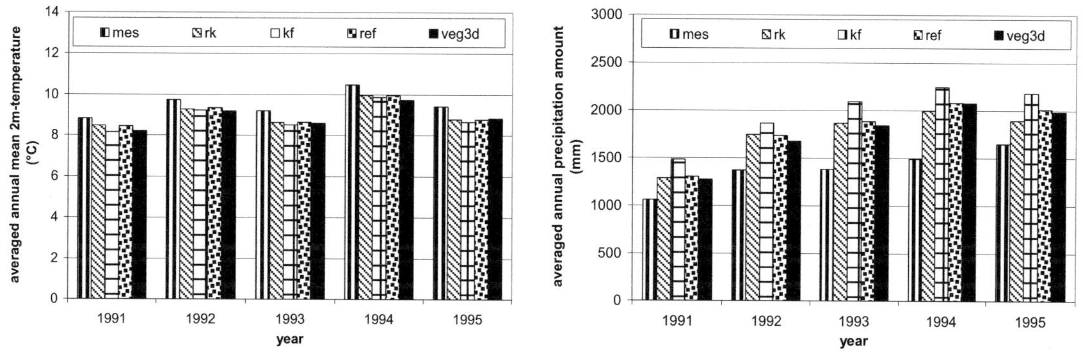

Figure 6.25: Station average of annual mean 2m-temperature (left) and annual precipitation sum (right) for the period 1991–1995 for observations and simulations (abbreviations are explained in the text).

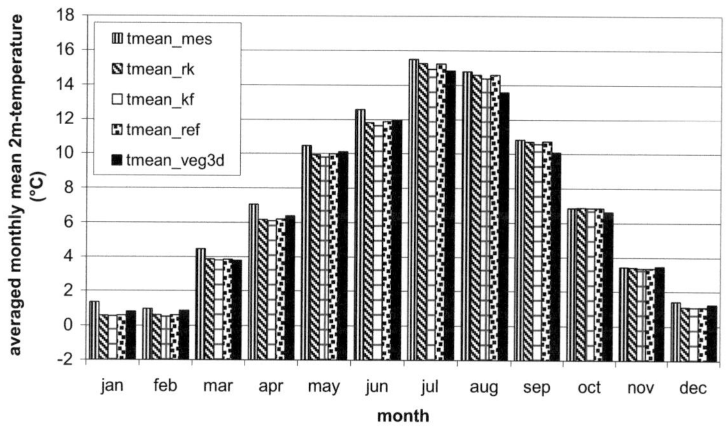

Figure 6.26: Station average of monthly mean 2m-temperature over the period 1991–1995 for observations and simulations (abbreviations are explained in the text).

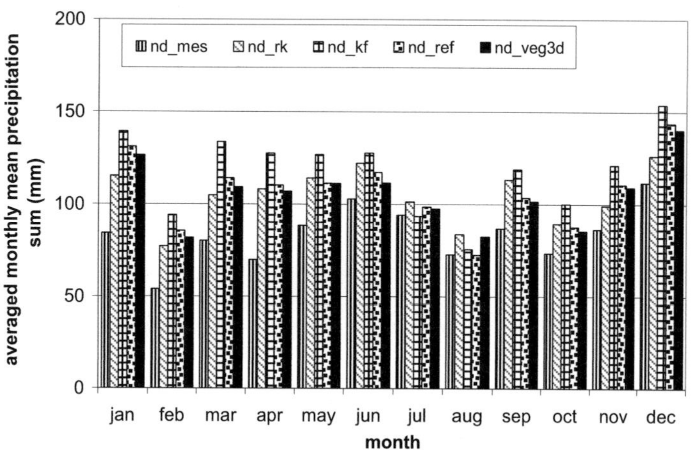

Figure 6.27: Station average of monthly mean precipitation sum averaged over the period 1991–1995 for observations and simulations (abbreviations are explained in the text).

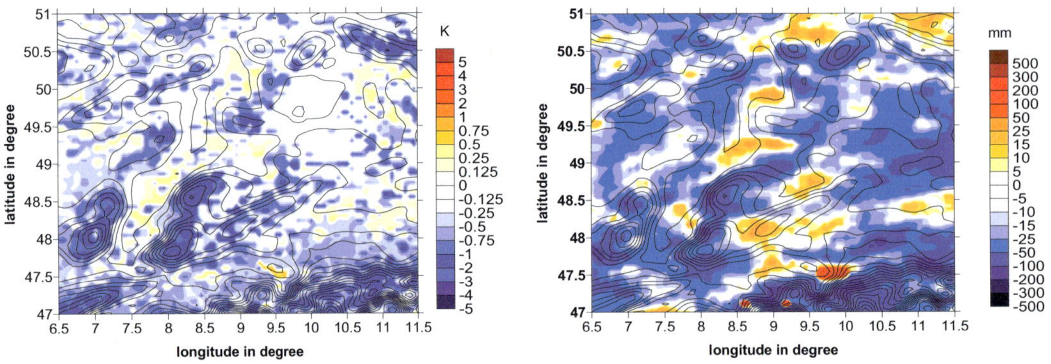

Figure 6.28: Difference of annual mean 2m-temperature (left) and annual precipitation sum (right) averaged over the period 1991–1995 between VEG3D and TERRA simulation.

The differences in annual mean 2m-temperature between the VEG3D simulation and the TERRA simulation are lowest for the Rhine valley and the area north of the Swabian Jura and are most pronounced over the forested areas and higher elevated sites (Fig. 6.28). The difference in the annual precipitation sum between VEG3D simulation and the TERRA simulation varies greatly, from -100 mm over the Black Forest to +50 mm over the Swabian Jura (Fig. 6.28). The area of strongest decrease of precipitation amount over the Black Forest can also be observed in the comparison for the year 2001.

6.5.2 Evaluation for the period from 1991 to 2000

The same evaluations are undertaken for the average of the annual values over the period 1991 to 2000 as they were for the year 2001. In general, the patterns of increase and decrease of precipitation and evapotranspiration in the VEG3D simulation, compared to the TERRA simulation, are the same as for the year 2001. The differences between the TERRA run and the VEG3D run are also of the same magnitude as they were for 2001 (Fig. 6.29, Fig. 6.30). The total precipitation amount decreases by about 100 mm on the western site of the Black Forest in the VEG3D run and there is a decrease in grid scale visible over the whole domain. The convective precipitation decreases at the western side of the Black Forest and the Swabian Jura and increases in the rest of the domain in the VEG3D simulation compared to the TERRA simulation. The large evaporation amount in the TERRA_LM simulation in the transition zones is also found for the decadal simulation.

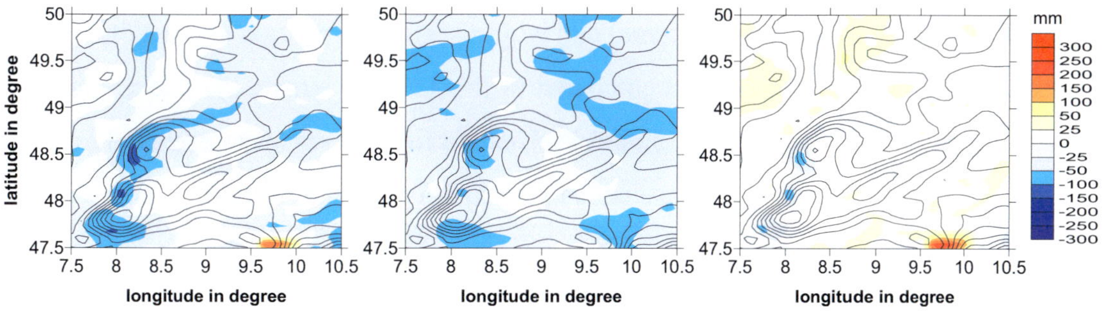

Figure 6.29: Averaged difference of annual precipitation sum between VEG3D and TERRA simulation for the period 1991–2000 for total precipitation (left), grid scale precipitation (middle) and convective precipitation (right).

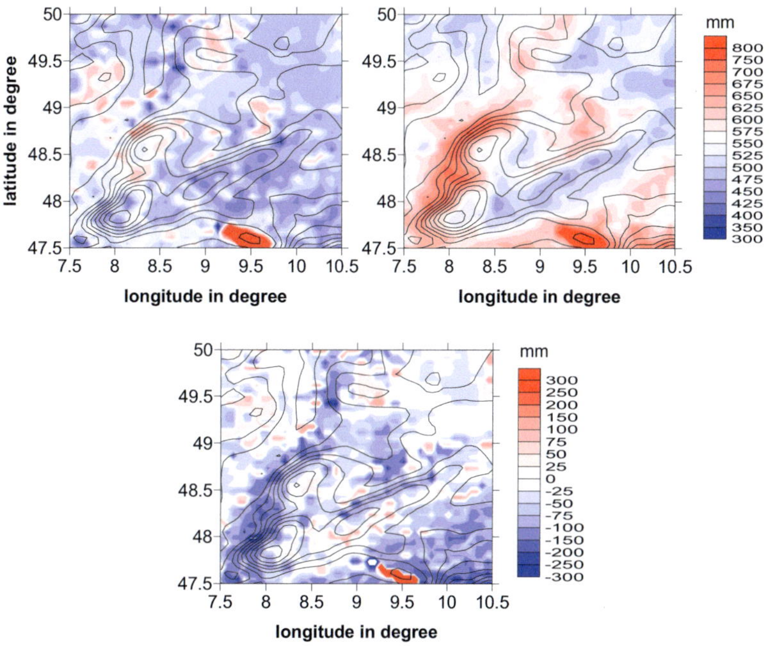

Figure 6.30: Annual evapotranspiration sum for VEG3D simulation (top, left) and TERRA simulation (top, right) and difference between VEG3D and TERRA annual evapotranspiration sum (bottom) averaged over the period 1991–2000.

Precipitation (mm)	**TERRA**	**VEG3D**
Total	1166	1136
Grid scale	829	774
Convective	347	362

Table 6.8: Area average of total, grid scale and convective precipitation for TERRA and VEG3D simulation averaged over the period 1991–2000.

A decrease in total precipitation (30 mm) and in grid scale precipitation amount (45 mm) is observed in the VEG3D simulation compared to the TERRA simulation and a slight increase of convective precipitation (15 mm) is observed for the area mean averaged over the period from 1991 to 2000. This corresponds to the findings for the year 2001.

The annual cycle of soil water content for the uppermost seven soil layers down to a depth of 1.40 m shows similar values for both simulations over the decade, especially during winter (Fig. 6.31). The drying out in summer in the uppermost layers in the TERRA run is visible throughout the whole decade. This is mainly due to the thinner root zone. In TERRA_LM, the root zone is restricted to the uppermost 0.8 m of the soil and therefore, water is mainly evaporated from the uppermost first meter of the soil, whereas in VEG3D, the root depth reaches much deeper into the soil. Water is also taken up from much deeper soil layers and the decrease in water content is not as pronounced in the VEG3D simulation as in the TERRA simulation for the uppermost 1.40m.

In the simulation with TERRA_LM the number of active layers is restricted to the uppermost seven soil layers and therefore, the total water content over all ten soil layers can only be evaluated for the VEG3D simulation. A comparison of the soil water amount of all soil layers where the soil water content is calculated prognostically shows that the soil water content in the TERRA run remains quite stable over the decade, only showing a decrease due to evapotranspiration in summer. The water content of the VEG3D simulation shows an increase over the first five years and remains nearly constant afterwards (Fig. 6.32). This can be explained by the fact that the initialisation is done by using a climatology produced with the TERRA_LM (see section 4.1) and therefore gives initialisation values for the soil moisture profiles that are well suited for the TERRA simulation but are slightly to low for the equilibrium state of the VEG3D simulation. The soil is filling up during the first five years until it reaches an equilibrium state.

The difference in monthly mean soil temperature for the uppermost seven soil layers between the VEG3D simulation and the TERRA simulation averaged over the investigation area shows an annual cycle. The soil temperatures in the summer months are about 1 K higher in the VEG3D simulation than in the TERRA run and lower or equal in winter (Fig. 6.33). This difference varies over the investigation area. As shown in Fig. 6.34, the main differences between both simulations occur in the Rhine Valley and north of the

Black Forest and the Swabian Jura. Over forested areas, such as the Black Forest, the soil temperature between both simulations is quite similar in summer and winter. Highest differences are obtained in summer for grassland and cultivated areas. In general, the differences over the whole investigation area are smaller in winter.

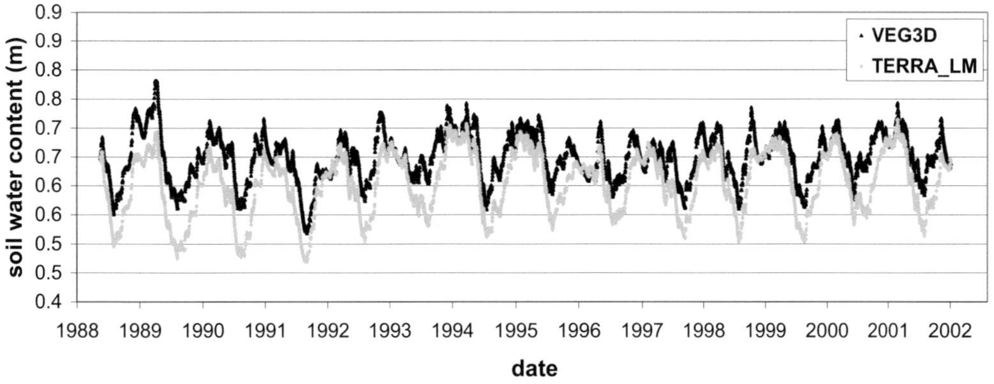

Figure 6.31: Annual cycle for mean total water content of the uppermost seven soil layers averaged over the investigation area.

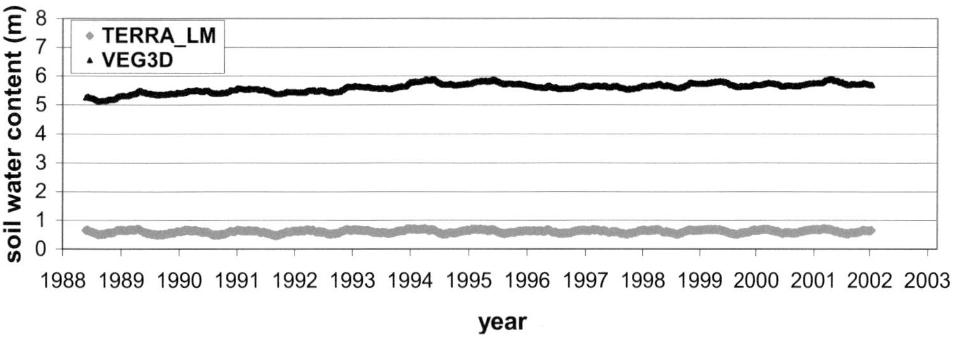

Figure 6.32: Annual cycle for the total water content in all active soil layers averaged over the investigation area.

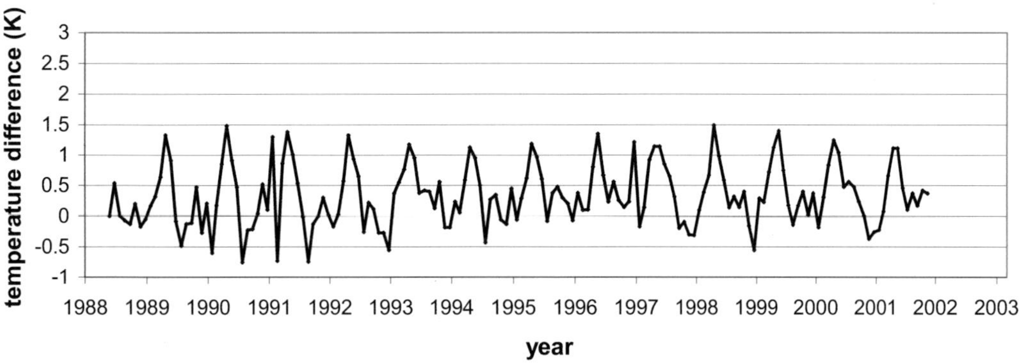

Figure 6.33: Annual cycle of the temperature difference between VEG3D simulation and TERRA simulation in the uppermost seven soil layers averaged over the investigation area.

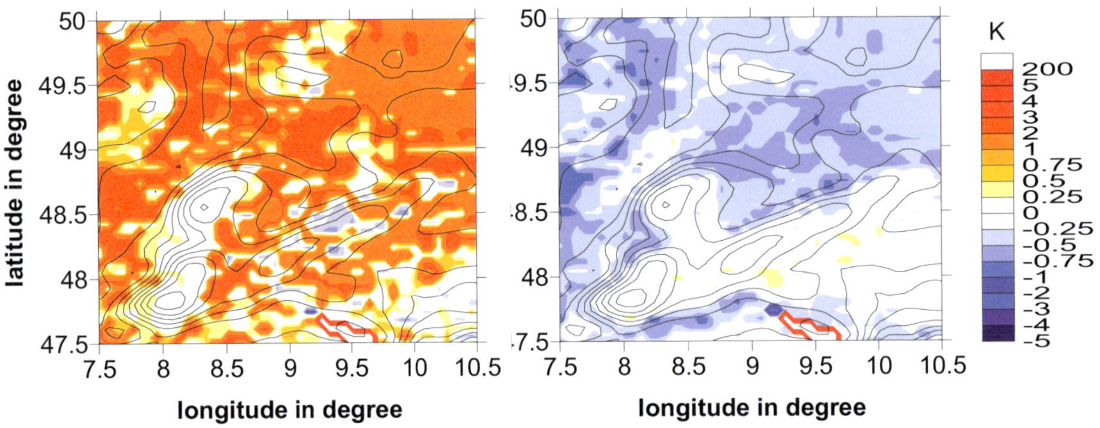

Figure 6.34: Difference in monthly mean soil temperature averaged over the uppermost seven soil layers between VEG3D simulation and TERRA simulation for June 1997 and December 1997.

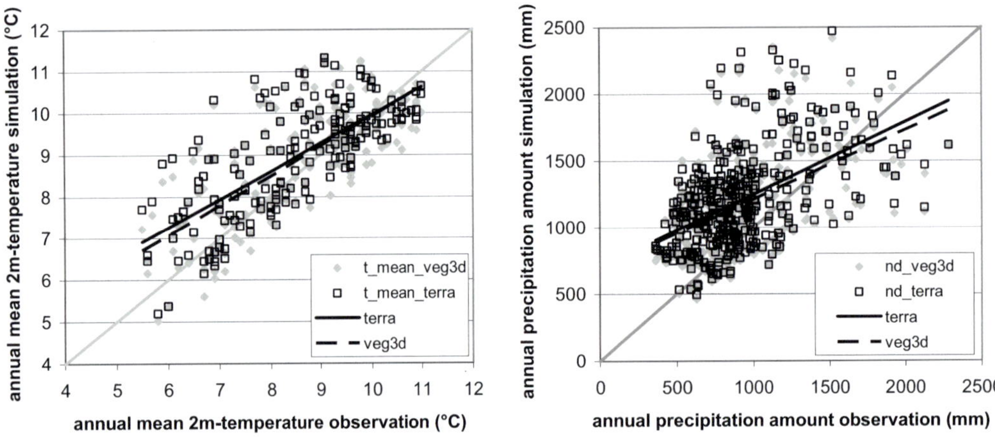

Figure 6.35: Observed versus simulated annual mean 2m-temperature and annual precipitation amount. Lines indicate the least squares fit and the grey line is the bisecting line.

Model run	Evaporation (sum)	Surface runoff (sum)	Ground ruoff (sum)	Precipitation (sum)	Water content (difference)
TERRA	5678	1933	4114	11643	37
VEG3D	5209	2344	3651	11355	219

Table 6.9: Water balance in millimetres averaged over all land points within the investigation area for the period from January 1991 to January 2001.

The linear fit between observations and simulations for annual mean 2m-temperature is quite similar in both simulations (Fig. 6.35). The correlation coefficient is 0.53 for the TERRA run and 0.52 for the VEG3D run. The absolute differences between the TERRA run and the VEG3D run can reach up 0.5 K at single stations.

The correlation coefficient is quite poor for the annual precipitation amount, with 0.26 for the TERRA run and 0.25 for the VEG3D run. The absolute differences between the reference run and the VEG3D run can reach 150 mm at single stations (Fig. 6.35).

The water balance was calculated for the period from January 1991 to January 2001 (Tab. 6.9). The evaporation and precipitation amount is quite similar in both simulations. The higher precipitation amount in the TERRA run is compensated by higher evaporation. The VEG3D simulation produces more surface runoff than the reference run (400 mm) but less ground runoff (450 mm). The difference in water content is 150 mm higher in the VEG3D simulation. In both models, an increase of soil water content is observed over the decade.

6.6 SUMMARY

Comparisons of simulations with the stand-alone versions of the schemes for a grassland and a forest site show increasing differences in soil temperature with increasing vegetation height. The damping effect of vegetation with regard to the amplitude of soil temperature is clearly visible. The partitioning into sensible and latent heat flux shows large differences between the two models. Overall, the agreement of the stand-alone simulations with observations is better for the grassland site than for the forest site. In order to improve the model performance over forested areas, better parameterisations are necessary, as well as more representative observations especially for the heat fluxes.

Comparisons of annual and decadal high-resolution simulations with the two different land surface schemes coupled online to the COSMO-CLM show a considerable influence of the land surface scheme on the simulation results of COSMO-CLM. The area average of the changes in annual precipitation amount and evapotranspiration is in the order of 5% and

10 % of the total amount and for several regions changes up to 10% and 20% occur. The highest differences in evapotranspiration can be observed in heterogeneous regions where different land use classes are contained within one model grid box. The differences between the two simulations in annual convective and grid scale precipitation reach up to 40% and 15% of the total amount respectively and are most pronounced at the windward side of the elevated sites. The area means of evaporation and total precipitation are higher for the TERRA_LM simulation than for the VEG3D simulation while convective precipitation is lower. Only small differences (0.5 K maximum) in the mean near-surface temperature between both simulations for non-forest stations can be found. The lower precipitation amounts in the VEG3D simulation accord better with the observed values.

The soil water content and the soil temperature trend show no pronounced differences between the simulations with the two land surface schemes. Nevertheless, seasonal differences between both simulations occur. Unlike the stand-alone versions of the models, the differences in soil temperature between the VEG3D and the TERRA simulation are greatest in areas with low vegetation and smallest in areas with high vegetation. This can be explained by the feedback mechanism with the atmosphere, which is not considered in the stand-alone simulations.

The differences in 2m-temperature and precipitation compared to the TERRA simulation are of the same magnitude as the changes in other physical parameterisations.

7 STATISTICAL-DYNAMICAL DOWNSCALING

Dynamical regional high-resolution climate simulations are very time consuming. The simulation of a 30 year period, which is a standard time period for climate investigations, may need weeks or even months of CPU computing time, depending on the computer platform available. Therefore, methods are required that can replace this method of explicitly simulating each single day. One possibility is statistical-dynamical downscaling, which consists of an algorithm classifying weather patterns from synoptic meteorological fields (e.g. reanalysis data), and then explicitly simulating the classified weather patterns (often corresponding to single days) with a regional model like COSMO-CLM. In this thesis, the classification is done by the SOM algorithm (Kohonen et al., 1995). In single-day simulations no adjustment time for the soil moisture and soil temperature profiles is available and a method is needed to provide adequate soil profiles for these simulations.

In the first part of this chapter the method of producing soil moisture and soil temperature profiles for single day simulations is presented and their influence on COSMO-CLM simulations is described. The second part presents a description of the SOM algorithm and a comparison of dynamical and statistical-dynamical simulations with observations.

7.1 INFLUENCE OF INITIAL SOIL MOISTURE AND SOIL TEMPERATURE PROFILES ON COSMO-CLM SIMULATIONS

The influence of soil moisture and soil temperature initialisation on long term RCM simulations has been investigated in section 4.2 and 4.3, and it has been shown that it takes at least one year of spin-up time for the soil profiles for the area of Southwest Germany if the initialisation is taken over from the coarse driving data. Therefore initializing the land surface scheme (LSS) in a regional model from the driving data with coarser resolution and different soil model physics for a daily simulation can cause unrealistic drifts in the model results due to the different equilibrium states in the driving model soil physics and the regional model soil physics (RODELL, 2005). The best method to provide adjusted initialisation data for the LSS of a regional model is therefore to start the continuous simulation with the regional model some years before the investigation period. This is rather time consuming and not practicable for statistical-dynamical downscaling because this spin-up time would destroy the advantage of saving time by only modelling single days. A practicable method for the initialisation of the LSS in short-term simulations with a regional model is the use of soil variable profiles produced by a stand-alone version of a LSS initialised some years before the starting day of the regional model simulation and run until this starting day (SMITH et al., 1994), ideally driven by the same kind of coarse driving data used for the regional model simulation later on and, for consistency, using the same LSS. The profiles of the soil variables have then some years to adapt to the physics of the LSS and these adjusted soil profiles can then be used to initialise the regional model. The computational costs are strongly reduced compared to the dynamical simulation with the complete regional model including the atmospheric part.

This initialisation method and its influence on regional climate simulations with COSMO-CLM have been investigated for the region of Southwest Germany. As driving data for the LSS and the regional model we used the ERA-40 reanalysis data (SIMMONS and GIBSON, 2000). A detailed description of the soil-vegetation model VEG3D, which is used here in stand-alone mode to produce the adjusted soil profiles, is given in section 6.1.2 and a description of the parameterisations used in COSMO-CLM is given in section 4.1. The simulation with the stand-alone version of VEG3D was started in 1995 and the COSMO-CLM simulation (version 3.22) took the soil variables from this simulation on the first of April 2001 as initialisation for the regional simulation. The regional model simulation was then run from the first of April to end of December 2001 with a horizontal grid resolution of 7 km and 64x64 grid points (Fig. 7.1). As driving data for the 7 km simulation a 50 km run with COSMO-CLM driven by ERA-40 reanalysis data for the whole European region was used (Fig. 7.1). Two 7 km runs were performed: one with the soil initialisation taken over from the ERA-40 reanalysis data (henceforth called ERA run) and one with the soil initialisation taken over from the run with the stand-alone soil model (henceforth called VEG3D run). Daily means and sums of observed meteorological variables for comparison with the simulations are available for 23 stations within the area (Fig. 7.1). The observations were provided by the German Weather Service (ÖSTERLE et al., 2006).

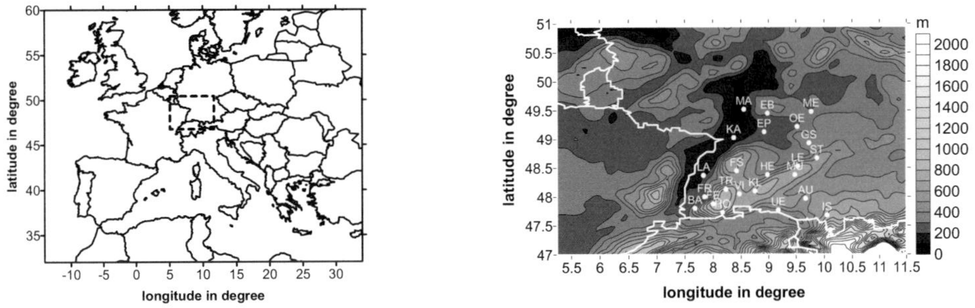

Figure 7.1: Left: Area of the 50 km COSMO-CLM simulation containing the area of the 7 km run (dashed rectangle). Right: Area of the 7 km run with observation stations indicated by white dots.

In the follwing root mean square errors (rmse) and root mean square deviations (rmsd) are calculated. The term error is used when simulations are compared to observations and the term deviation is used when two simulations are compared.

The soil moisture and soil temperature profiles are quite different in the two model runs at the initialisation day. We find soil temperature differences of up to 6.5 K and differences of up to 7 volume percent in soil moisture content for the area average of the whole investigation area (Fig. 7.2). The rmsd of soil moisture content between the two simulations decreases considerably for the whole profile during the first two months and stays nearly constant during the rest of simulation (Fig. 7.3). The rmsd of soil temperature shows a strong decrease only in the upper soil levels during the simulation. In the lower levels the temperature cannot adapt because the temperature at the lowest level is kept constant on the initialisation temperature over the whole simulation (Fig. 7.3). At the end of the simulation in December 2001 the soil moisture profile is rather similar in both simulations and the rmsd is less than one volume percent for the area average. The temperature profile has only slightly smoothed and shows increasing rmsd with increasing soil depth.

In the first month, the rmse between the observed and modelled monthly mean 2m-temperature and the monthly precipitation is largest for the 23 stations (Fig. 7.4). The rmse of 2m-temperature is about 0.4 K lower in the VEG3D run than in the ERA run and 2m-temperature is therefore better simulated by the VEG3D run. The rmse of precipitation increases in the VEG3D simulation by about 5 mm in the first month. After three months of simulation time we get quite similar results for the rmse of the VEG3D and the ERA simulation.

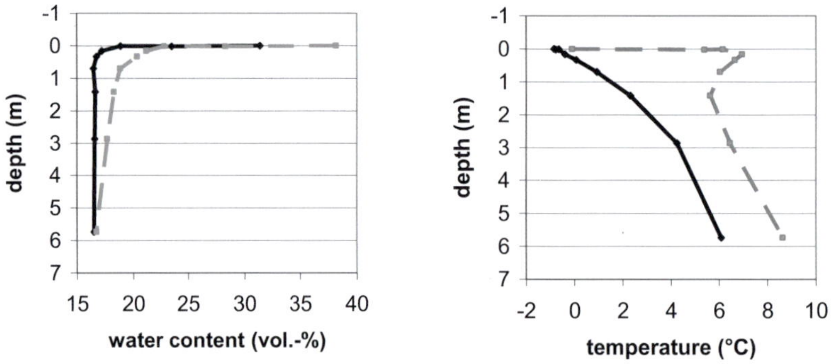

Figure 7.2: Profiles of soil moisture content (left) and soil temperature (right) averaged over the whole investigation area at the initialisation date for ERA (black solid line) and VEG3D (grey dashed line).

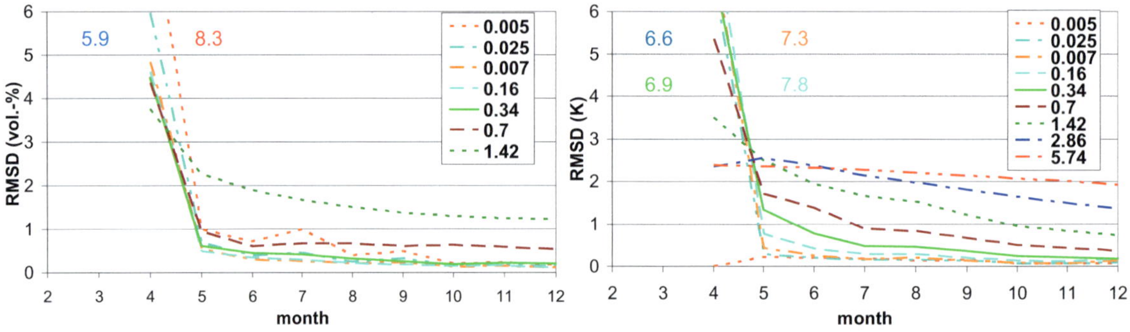

Figure 7.3: Root mean square deviation of soil moisture content (left) and soil temperature (right) between the two simulations for different soil depths (in metres) for the simulation period from April to December 2001.

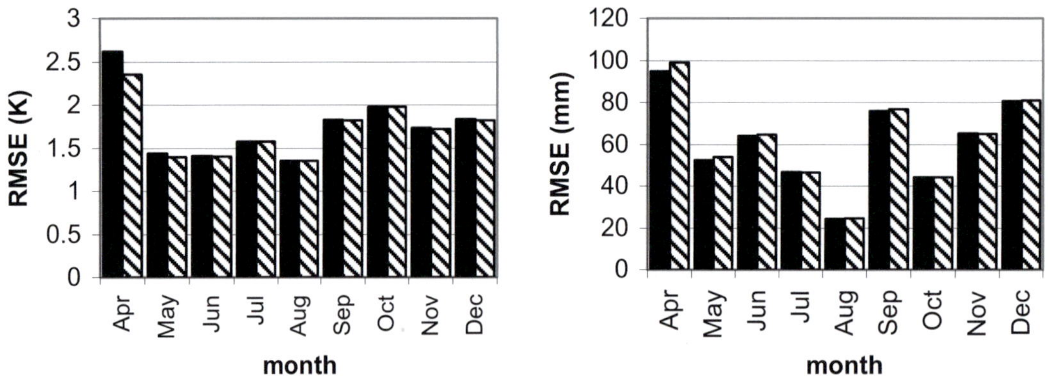

Figure 7.4: Root mean square error between monthly mean 2m-temperature (left) and precipitation sum (right) of observations and ERA (black) and observations and VEG3D (shaded) for the simulation period from April to December 2001.

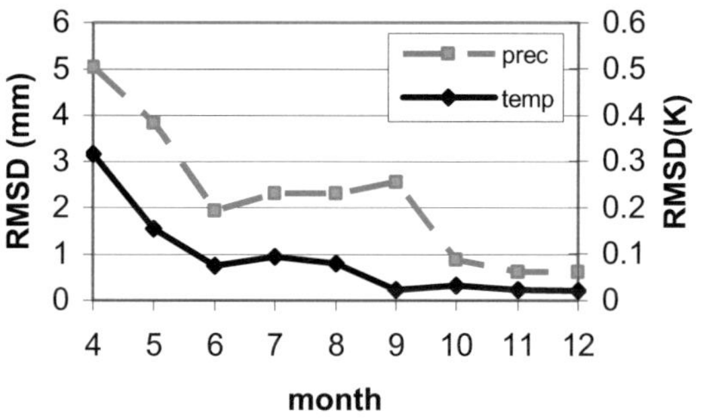

Figure 7.5: Root mean square deviation between ERA and VEG3D run for monthly precipitation sum (dashed line) and monthly mean 2m-temperature (solid line).

The rmsd between both simulations for the monthly mean and sum of 2m-temperature and precipitation approaches zero over the simulation time (Fig. 7.5). After a strong decrease during the first two months the rmsd stays constant for 2m-temperature and even increases for precipitation during the summer months from June to August. This is due to the convection during the summer months, which is influenced even by small changes in soil moisture and soil temperature and has a strong non-linear feedback on precipitation amount and 2m-temperature, which prevents the rmsd decreasing further. In the late autumn, the precipitation is mainly grid scale and the influence on the precipitation amount by variations in soil moisture and soil temperature decrease so that the RMSD of precipitation decreases further.

The rmse of daily mean 2m-temperature between observation and simulation is strongly reduced in the VEG3D simulation for the first day of simulation compared to the ERA run by up to 2 K (Fig. 7.6). The rmse for daily precipitation is slightly enhanced in the VEG3D run in the first days of simulation by a maximum of 0.5 mm (Fig. 7.6). This is caused by the higher evapotranspiration due to the higher soil moisture content in the upper soil levels.

The initialisation of the soil moisture and soil temperature profiles has a strong impact on the results of the temperature forecast for the first three days of the simulations, with differences of at least 0.5 K in the rmse between the two simulations. The rmse shows the same values in precipitation for both simulations for the first time after 13 days and after this period the difference of rmse for 2m-temperature is lower than 0.1 K for the first time.

Comparing the precipitation amount for the whole simulation period, differences of up to 50 mm occur between both simulations with an enhancement of precipitation in the southeast of the investigation area and a decrease in the northwest for the VEG3D run (Fig. 7.7). This increase in precipitation corresponds to the increase in evapotranspiration in the VEG3D run for the southern area, which is up to 100 mm higher than in the ERA simulation (Fig. 7.7).

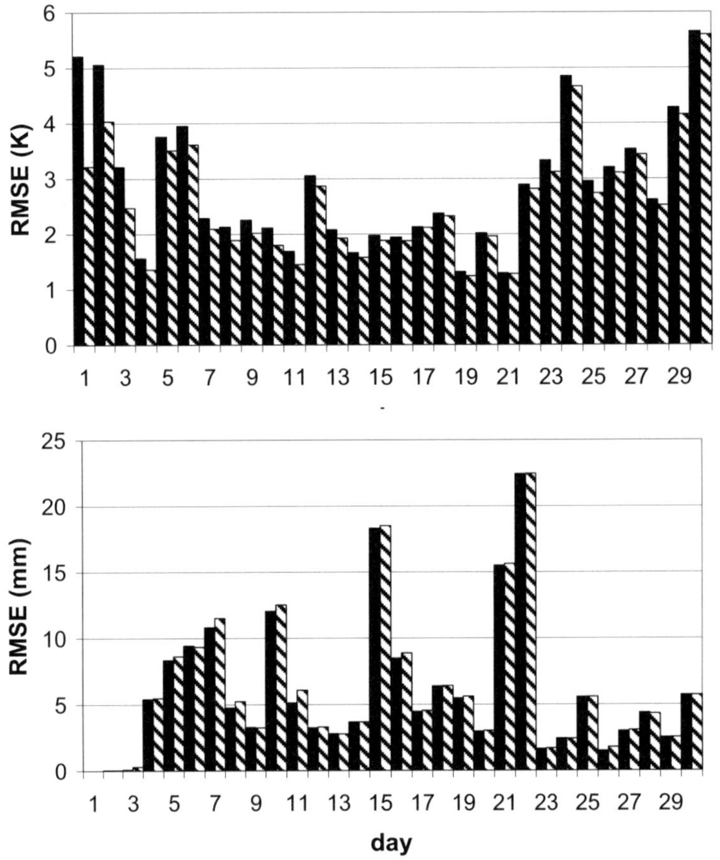

Figure 7.6: Root mean square error between daily mean 2m-temperature and daily precipitation sum of observations and ERA (black) and observations and VEG3D (shaded) simulation for 2m-temperature (top) and precipitation (bottom) for April 2001.

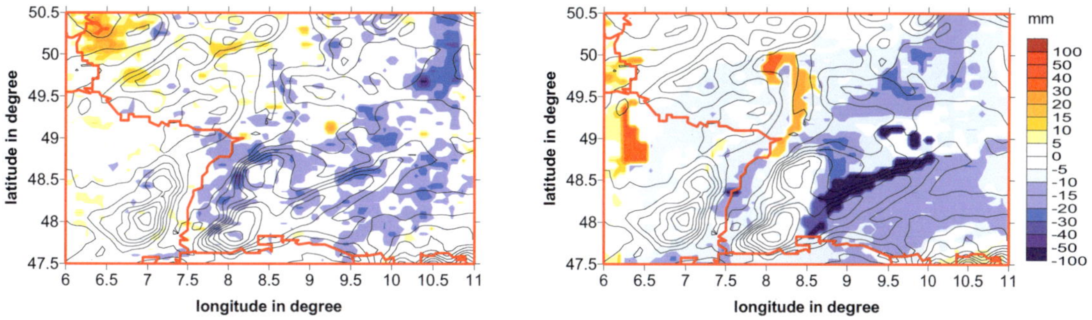

Figure 7.7: Difference between ERA and VEG3D simulation for the total precipitation amount (left) and the evapotranspiration (right) amount for the period April to December 2001.

Using soil moisture and soil temperature profiles pre-processed by a stand-alone soil model seems to be an appropriate solution for the initialisation problem occurring for short-term simulations. In this case the use of pre-processed soil profiles improves the 2m-temperature forecast considerably for the first days of simulation. The higher soil moisture content, which leads to an unrealistic enhancement of precipitation especially in the south-east of the investigation area, may have several causes: (1) the driving data of the stand-alone model produce too high precipitation amounts for this region so that the soil is too wet at this initialisation date or (2) the equilibrium soil moisture profile provided by the stand-alone model of VEG3D is not consistent with the equilibrium soil moisture profile that would be produced by the soil model TERRA_LM used in the COSMO-CLM, and the surplus of soil moisture is therefore released into the atmosphere by evapotranspiration. To determine which of these effects is the cause for the high precipitation amount, additional comparisons with high-resolution observation data would be necessary. This is difficult due to the lack of adequate observations of soil moisture, soil temperature and heat flux measurements.

7.2 Comparison of dynamical and statistical-dynamical downscaling for the year 2001

7.2.1 Using Self-Organizing Maps for statistical-dynamical downscaling

Self-organizing maps (SOM) (Kohonen, 1995) are a kind of cluster analysis used here to classify weather patterns by using fields of meteorological variables. A detailed description of the algorithm and its application to meteorological fields can be found in Sasse (2004) and therefore only a short description of the algorithm is given here:

Using n two-dimensional fields of meteorological variables with nx longitudinal grid points and ny latitudinal grid points a ($n*nx*ny$)-phase space of input vectors is created for which a prescribed number of nodes are defined and distributed randomly in this phase space. The position of the nodes is then iteratively adjusted by their Euclidean distance to the $n*nx*ny$ input vectors. During one iteration cycle, not only the actually considered node is drawn towards an input vector, but also other nodes in a defined neighbourhood are drawn towards the input vector. At the end of the iterations a certain number of input vectors can be assigned to each node, which are closer to the specified node than to any other node in the phase space. The ratio of the number of input vectors assigned to one node and the total number of input vectors then gives the relative frequency of the node.

There are two major differences between the SOM methodology and other clustering algorithms (Hewitson and Crane, 2002):

(1) The primary goal of the algorithm is not grouping data or identifying clusters. Instead, SOM attempt to find positions of the nodes in the data space that are representative of the nearby clouds of data and describe the multi-dimensional distribution function of the data set all together. Thereby nodes are clustered in regions with high data density.

(2) More than one node is updated during the iteration cycle: Surrounding nodes are incrementally adjusted depending on their distance to the node considered.

An advantage of SOM is that there is no statistical model needed, which describes the data distribution. Intensive studies with geopotential and surface pressure fields have been carried out by SASSE (2004) to determine the parameters adjustable in the algorithm, e.g. the number of iteration steps, the number of nodes and the neighbourhood function. He found that the number of iteration steps should be 100.000 and that 20 nodes seem to be an adequate number for classifying near-surface pressure fields and that a step-neighbourhood function for the determination of the influenced nodes is better than a gaussian one. Sasse also found that the size of the meteorological fields is very important: The area has to contain the typical structures of the investigation area (SCHÄDLER and SASSE, 2006).

For this study the classification of 20 nodes (weather patterns) and their relative frequency for the year 2001 obtained by HALLER (2005) is used for the statistical-dynamical downscaling: the classification has been done by using the geopotential in 500 hPa and/or the surface pressure. Every class represents one characteristic weather pattern and is represented by one day of the year. This day is simulated explicitly with COSMO-CLM and the daily means of meteorological variables are then multiplied by the relative frequency of the class. This is done for all classes and the sum of this product of daily mean times frequency is then divided by the total number of days in the year and the annual mean of the meteorological variables is obtained.

Unlike the method used by SASSE (2004), who simply multiplied the surface pressure by a constant factor to get geopotential and surface pressure in the same range for the SOM algorithm, the method presented by HALLER (2005) normalises the meteorological fields by:

$$x' = \frac{x - \overline{x}}{\sigma}.$$

The normalised field x' is obtained by subtracting the area mean of the field $\overline{x}$ from the input field x and dividing this difference by the standard deviation σ of the field.

For the classification of the 20 classes, HALLER (2005) used NCEP reanalysis data (KANAMITSU et al., 2002) for the area from 30° north to 70° north and from 20° west to 30° east.

7.2.2 Comparison of the two different downscaling techniques for the year 2001

Gerstengarbe et al. (1999) classified 29 synoptic weather patterns ("Grosswetterlagen") for Central Europe from observed meteorological data. From these patterns, nine of them have a relative frequency of less than 2% so that a number of 20 classes for one year for the SOM algorithm seems adequate to allow the algorithm to classify the most common weather patterns. Four different combinations of geopotential in 500 hPa and surface pressure have been used by Haller (2005) for the classification:

- Geopotential 500 hPa without normalisation (Gp)
- Geopotential 500 hPa with normalisation (Gpn)
- Geopotential 500 hPa and surface pressure without normalisation (GpSp)
- Geopotential 500 hPa and surface pressure with normalisation (GpSpn)

The 20 classes, with their relative frequency for the year 2001, are shown in Tab. 7.1. The algorithm does not classify all of the 20 Großwetterlagen for that year. As shown by Haller (2005), some days classified by SOM belong to the same Großwetterlage. Depending on the parameter used for classification different days of the year are classified and therefore simulated explicitly by the climate model for the statistical-dynamical downscaling.

As is obvious from Tab. 7.1, the algorithm classifies mainly days in spring, autumn, and winter. Summer days are rare especially for the classes GpSp and Gp. This means that convective summer days, which are quite often observed in the area of Southwest Germany, are not well represented in 2001 although they contribute an important amount of convective precipitation for that region in summer in many other years. This means also that the number of warm days and their frequency, which are important for a correct 2m-temperature, is smaller than the number of cold days or days with moderate temperature.

From the diploma thesis of Haller (2005), it is known that the statistical-dynamical downscaling yields good results for temperature compared to observed data. But the strategy used by Haller is not applicable for climate simulations: he simulated the year 2001 in a continuous dynamical simulation and extracted from this run the 20 days classified by the

SOM algorithm. With these 20 days he calculated annual means by using the relative frequency given by SOM. This strategy cannot be used for climate runs because the dynamical simulation of decades is to be avoided in the statistical-dynamical downscaling. The strategy presented here is therefore different. The 20 days are simulated individually by COSMO-CLM as single day simulations. The daily means and sums of meteorological variables from these 20 simulations are multiplied by the relative frequency given by the SOM algorithm and then summed up to annual means.

SOM	**GpSpn**	**GpSp**	**Gpn**	**Gp**
class	**Day freq.**	**Day freq.**	**Day freq.**	**Day freq.**
1	12.12. 4.93%	28.10. 4.11%	**09.08.** 4.11%	01.03. 2.19%
2	17.02. 4.66%	01.04. 3.56%	12.07. 4.66%	04.01. 6.21%
3	18.11. 3.56%	05.04. 4.38%	03.01. 6.85%	05.04. 4.38%
4	**01.06**. 6.58%	04.01. 6.21%	**21.08.** 6.03%	01.04. 3.56%
5	16.04. 6.75%	01.03. 2.19%	**03.08.** 6.75%	28.10. 4.11%
6	**22.06.** 7.67%	30.05. 9.86%	05.03. 2.74%	25.12. 4.93%
7	**27.07.** 4.11%	05.10. 6.75%	07.04. 3.56%	19.03. 6.75%
8	06.11. 4.11%	23.04. 3.29%	09.04. 7.40%	22.01. 3.29%
9	08.04. 4.93%	19.03. 6.75%	27.07. 4.66%	05.10. 6.75%
10	26.02. 2.74%	27.12. 4.93%	23.10. 4.93%	03.09. 9.86%
11	17.01. 3.84%	**09.08.** 6.03%	20.03. 6.03%	05.11. 2.47%
12	**24.07.** 7.12%	21.10. 4.93%	02.09. 3.56%	20.04. 2.74%
13	**07.06.** 6.21%	17.01. 3.56%	06.11. 7.4%	03.05. 3.56%
14	13.03. 3.56%	20.04. 2.74%	16.02. 4.38%	21.10. 4.93%
15	18.03. 4.93%	06.11. 2.47%	18.01. 4.93%	**09.08.** 6.03%
16	06.10. 4.66%	**30.06.** 10.68%	26.12. 6.21%	31.01. 4.11%
17	14.10. 6.58%	21.09. 6.58%	**03.06.** 4.38%	14.02. 4.93&
18	11.03. 6.48%	10.12. 4.93%	31.01. 6.03%	10.12. 4.93%
19	03.01. 4.93%	14.02. 4.66%	12.01. 2.74%	21.09. 6.58%
20	24.03. 4.66%	31.01. 4.38%	10.12. 4.66 %	**30.06.** 10.68%

Table 7.1: SOM classes with assigned day and relative frequency for the four combinations of geopotential and surface pressure. Summer days are bold typed.

Due to the coarse resolution of the ERA-40 reanalysis driving data we have to use the same downscaling chain as explained in section 4.1: first the 50 km simulations driven by ERA-40 data for whole Europe have to be performed for these 20 days and then the daily high-resolution runs with 7 km horizontal resolution for the alps area are nested therein. The simulations have been done for all four combinations of geopotential and surface pressure, once with the initialisation of soil moisture and soil temperature profiles from the ERA-40 driving data and once from the pre-processed soil profiles obtained by VEG3D, as presented in the previous section (referred to in the following as OB and MB simulation). In total, we have eight simulations that can be compared to observations and the continuous ERA-40 simulation (ERA) with 7 km horizontal resolution evaluated in subsection 4.3.2.

Three main questions will be answered by this comparison:

(1) Is the method of statistical-dynamical downscaling able to reproduce the results obtained by the dynamical downscaling and by observations?

(2) Which combination of geopotential and surface pressure produces the most reliable results?

(3) Does the use of pre-processed soil moisture and soil temperature profiles enhance the results of statistical-dynamical downscaling?

In the following 2m-temperature and total precipitation were examined to answer these three questions. For comparison of the statistical-dynamical downscaling results with observations the data from the 23 DWD stations were used, as described in the section before.

To get an initial idea of the representativeness of the SOM classes for the year 2001, the observed annual mean 2m-temperature and the annual total precipitation amount at the 23 weather stations are compared to the results for the statistical-dynamical method using observed daily means and sums instead of values from the COSMO-CLM simulation to calculate the annual mean value and the annual sum of 2m-temperature and precipitation. By using observations as "perfect" simulations, the skill of the method and the classified days can be determined. Averaging the annual mean 2m-temperature and the annual precipitation sum over the 23 stations we get a mean value that can be used for a first evaluation of the reliability of the results. Using the statistical-dynamical method with observed data we get maximum differences in this mean value of up to 1.9 K for 2m-temperature and 350 mm in precipitation between the single combinations (Tab. 7.2). The best results are obtained by using the GpSpn combination with a difference of only 0.1 K between the observed averaged annual mean 2m-temperature and the statistical-dynamically calculated one and of only 50 mm between the mean precipitation amounts for these two methods.

The difference for the single observation sites between continuously observed annual mean 2m-temperature and statistical-dynamically produced annual mean 2m-temperature with the combination GpSpn is lower than 0.4 K, which is quite good. The difference in precipitation amount between the observed and calculated values can reach 800 mm at single stations. For 14 sites the difference is more than 100 mm and at most of these sites precipitation is underestimated by the statistical-dynamical downscaling. This compensates for the strong overestimation of 822 mm for the Feldberg site so that the average over all sites shows quite good results for the statistical-dynamical method with GpSpn combination. The ranking of the combinations is the same for temperature and for precipitation: GpSpn gives the best results compared to continuous observations followed by Gp, GpSp and Gpn. This order was expected because the use of two instead of one meteorological field should improve the classification of adequate days and, therefore, the simulation results and the normalisation when using two meteorological fields should give better results than using two fields without normalisation.

The statistical-dynamical downscaling method does not match perfectly the continuous observations, but the 20 classes identified by the SOM algorithm seem to be placed adequately to represent the annual mean and sum at least for the GpSpn and Gp combination. But the strong differences in annual precipitation obtained with even “perfect” conditions should be kept in mind for the evaluation of the statistical-dynamical method used with simulations later on.

Station	obsT	GpSpnT	GpSpT	GpnT	GpT	obsP	GpSpnP	GpSpP	GpnP	GpP
AULEND	8.8	8.9	8.39	6.98	8.44	916.5	818.8	918.5	1683.8	806.6
BADENW	10.1	9.9	9.81	8.07	9.98	1240.3	1197.4	1035.9	1364.2	720.1
EBERBA	10.1	9.9	9.13	8.04	9.59	1267.2	1229.7	768.7	1698.8	1117.1
EPPING	10.5	10.4	9.7	8.62	10.13	924.2	845.4	511.4	979.5	882.1
FELDBE	3.6	3.4	2.9	1.64	3.09	1944.1	2766.9	1111.1	2487.3	727.7
FRHBF	11.8	11.6	11.3	9.87	11.58	1127.6	1438.7	1308.9	1348.2	705.6
FST	7.3	7.3	6.9	5.32	6.99	1887.7	1817.1	1024.3	2671.7	1098.1
GSCHWE	8.9	8.7	8.3	7.1	8.74	1263.0	850.9	870.7	1300.2	724.4
HECHIN	9.1	8.8	8.5	7.04	8.72	916.2	671.5	311.3	1020.5	301.7
HOEHEI	9.3	8.9	8.4	7.41	8.75	1157.6	1188.3	1556.8	1435.4	1945.1
ISNY	7.5	7.7	7.2	5.96	7.57	1887.5	1768.4	2229.4	3165.5	2174.5
KARLSR	11.3	11.1	10.4	9.39	10.69	873.1	999.8	739.2	808.5	1355.1
KLIPPE	6.9	6.9	6.4	5.3	6.7	1147.4	892.5	416.9	1085.2	411.6
LAHR	11.0	10.8	10.4	9.04	10.58	1030.3	967.3	451.9	1328.9	336.4
LENNIN	8.3	8.2	7.8	6.2	8.1	1229.2	1116.5	820.9	1702.8	807
MANNHE	11.1	10.8	10.2	9.16	10.68	726.1	895.9	467.1	1250.2	554.6
MERGEN	9.7	9.7	8.8	7.82	9.43	938.6	755.5	561.8	931.5	870.2
MUENSI	7.4	7.3	6.7	5.35	7.06	1087.2	1032.8	684.2	1740.9	772.1
OEHRIN	10.1	10	9.2	8.18	9.79	940.3	544.4	582.9	735.4	918.3
STOETT	7.6	7.4	6.9	5.71	7.58	1223.5	836.5	645.2	1481.4	716.6
TRBG	7.8	7.6	7.2	6.1	7.35	2196.0	2323.6	1560.8	2248.6	1545.2
UEBERL	9.2	9.1	8.7	7.3	8.85	1107.3	1078.9	794	1810.8	815
VILLIN	7.5	7.3	7.0	5.55	7.16	1122.9	954.9	675.4	804.6	798.5
Mean	*8.9*	***8.8***	*8.3*	*7.0*	*8.6*	*1224.1*	***1173.6***	*871.6*	*1525.4*	*917.5*

Table 7.2: Results of the statistical-dynamical method using observed meteorological data for annual mean 2m-temperature (T) and annual precipitation sum (P) and the annual values obtained by continuous observation (obs) at the 23 stations.

7.2.2.1 COMPARISON OF THE RESULTS OF THE TWO DOWNSCALING SCHEMES FOR ANNUAL MEAN 2M-TEMPERATURE

HALLER (2005) found that for 2m-temperature best results compared to observations are obtained by using the GpSpn combination. This is also found with the statistical-dynamical method using simulations and with the statistical-dynamical method using observations. Comparing the average over all 23 sites we find a value of 8.9°C for the observations, 8.1°C for the dynamical simulation (ERA) and 8.0°C for the GpSpnOB simulation (Tab. 7.3). The reason for the too low temperatures of simulations compared to observations seems to be the ERA-40 reanalysis data (see section 4.2 and 4.3). The difference between dynamical and statistical-dynamical downscaling is rather small (0.1 K) for the GpSpnOB simulation but the absolute difference at single sites is quite large with up to 0.8 K. Using the VEG3D profiles for initialisation we also get best results for the GpSpn simulations but the mean temperature is 0.4 K lower than with initialisation of soil profiles from ERA-40 data. The absolute difference of the mean temperature over all sites between the four combinations of geopotential and pressure is up to 1.8 K, which is comparable to the range given by the statistical-dynamical downscaling with observed data. The lowest mean temperature, and therefore the worst result compared to observations, yields the Gpn combination for MB and OB simulation, in agreement with the results of HALLER. With 1.6 K the spread in the MB simulations is quite similar to the spread of the OB simulations. The ranking of simulation results is the same for OB and MB simulations: the best result compared to observations is obtained by the GpSpn simulation followed by the Gp simulation, the GpSp simulation and the Gpn simulation.

With the GpSpn combination, the statistical-dynamical downscaling is able to reproduce the dynamical downscaling results. For half of the sites we get better results for annual 2m-temperature with the statistical-dynamical method than with the dynamical method compared to observations.

To explain the lower 2m-temperatures in the MB simulations, the difference of OB and MB simulations was compared for the whole investigation area (Fig. 7.8). For the Rhine Valley, the MB simulations show higher absolute annual mean 2m-temperatures than the OB simulations and for the area west of the Rhine Valley and north of the Swabian Jura the annual mean temperatures of MB simulation is only up to 0.4 K lower than the

OB simulation for the GpSpn and GpSp combination and even up to 0.3 K higher than the OB simulations for the Gp and Gpn simulations. This means that, for some regions, the MB simulations are as warm as the OB simulations and in some regions even warmer - in better agreement, therefore, with observations.

Station name	Abbr.	Obs	ERA	GpSpn OB	GpSpn MB	GpSp OB	GpSp MB	Gpn OB	Gpn MB	Gp OB	Gp MB
AULENDORF-SPIEGLER	AU	8.8	8.3	7.5	7.0	6.6	6.5	5.7	5.4	7.0	6.7
BADENWEILER	BA	10.1	9.4	9.4	9.3	8.7	8.6	7.2	7.4	8.7	8.6
EBERBACH/NECKAR	EB	10.1	9.5	9.7	9.4	8.8	8.7	8.1	8.0	9.2	9.2
EPPINGEN	EP	10.5	9.8	9.4	9.4	8.6	8.7	7.8	7.9	8.9	9.1
FELDBERG/SCHW.	FE	3.6	2.5	2.9	2.6	2.5	2.3	1.1	1.0	2.4	2.2
FREIBURG I. BR. HBF	FR	11.8	10.3	9.6	9.8	8.9	9.1	7.6	7.9	9.0	9.1
FREUDENSTADT	FS	7.3	6.4	6.6	6.4	5.9	5.6	4.7	4.6	6.0	5.8
GSCHWEND KR. OSTALB	GS	8.9	8.0	8.0	7.7	7.3	7.2	6.4	6.2	7.8	7.7
HECHINGEN	HE	9.1	8.6	8.8	7.9	8.0	7.2	6.8	6.1	8.2	7.4
HOECHENSCHWAND	HO	9.3	5.6	5.8	5.4	5.0	4.9	3.7	3.7	5.1	4.9
ISNY	IS	7.5	7.5	6.9	6.5	6.2	6.2	5.1	5.2	6.6	6.5
KARLSRUHE	KA	11.3	10.8	9.6	9.9	8.6	9.1	7.7	8.2	8.9	9.4
KLIPPENECK	KL	6.9	5.8	6.0	5.2	5.5	4.7	4.1	3.6	5.5	4.8
LAHR	LA	11.0	10.4	9.8	9.9	9.0	9.1	7.8	8.0	9.1	9.2
LENNINGEN	LE	8.3	6.9	7.1	6.2	6.5	5.9	5.5	4.8	7.0	6.1
MANNHEIM	MA	11.1	10.2	9.7	9.8	8.9	9.2	8.0	8.3	9.3	9.5
MERGENTHEIM	ME	9.7	9.5	9.3	9.1	8.6	8.5	7.7	7.6	9.2	9.1
MUENSINGEN	MU	7.4	6.9	7.0	6.1	6.1	5.6	5.4	4.7	6.6	5.8
OEHRINGEN	OE	10.1	9.4	9.5	9.2	8.8	8.6	7.9	7.8	9.2	9.1
STOETTEN	ST	7.6	6.8	7.0	6.1	6.4	5.7	5.5	4.7	6.9	6.0
TRIBERG	TR	7.8	7.5	7.8	7.5	7.3	7.0	5.9	5.9	7.3	7.0
UEBERLINGEN/BODENSEE	UE	9.2	8.7	8.1	6.5	7.1	6.1	5.9	5.0	7.2	6.1
VILLINGEN-S.	VI	7.5	7.5	7.5	7.2	6.9	6.6	5.5	5.4	6.8	6.7
Mean		*8.9*	*8.1*	*8.0*	*7.6*	*7.3*	*7.0*	*6.2*	*6.0*	*7.5*	*7.3*

Table 7.3: Annual mean 2m-temperature in °C for observations (obs), the dynamical downscaling simulation (ERA) and the statistical-dynamical simulations with COSMO-CLM for the four combinations of geopotential and near-surface pressure with soil initialisation from ERA-40 data (OB) and from VEG3D profiles (MB) for the year 2001.

For the south eastern part of the investigation area, ranging from the Swabian Jura to the Alps, differences of up to 2 K occur between the MB and the OB simulations and these are strongest in the GpSpn combination. The rectangular structure of the area, which covers nearly one quarter of the investigation area, gives a hint that these strong differences come from the soil profiles obtained by the stand-alone LSS model driven by the ERA-40 reanalysis data. The investigation area is covered by 4 ERA-40 reanalysis grid points and the area covered by one ERA grid point covers the area of strongest differences exactly.

When the grid point data used to drive the LSS in this area is too cold or contains precipitation rates that are too high, the values of the soil temperature profiles of the LSS used for initialisation of the regional model can be too low and can cause the observed too low 2m-temperature in the MB simulations. To determine whether these too low temperatures are an effect observable over the whole year the daily means of 2m-temperature for the 20 classes are plotted for three sites within the south-eastern region: Aulendorf (Au), Hechingen (HE) and Überlingen (UE) (Fig. 7.9).

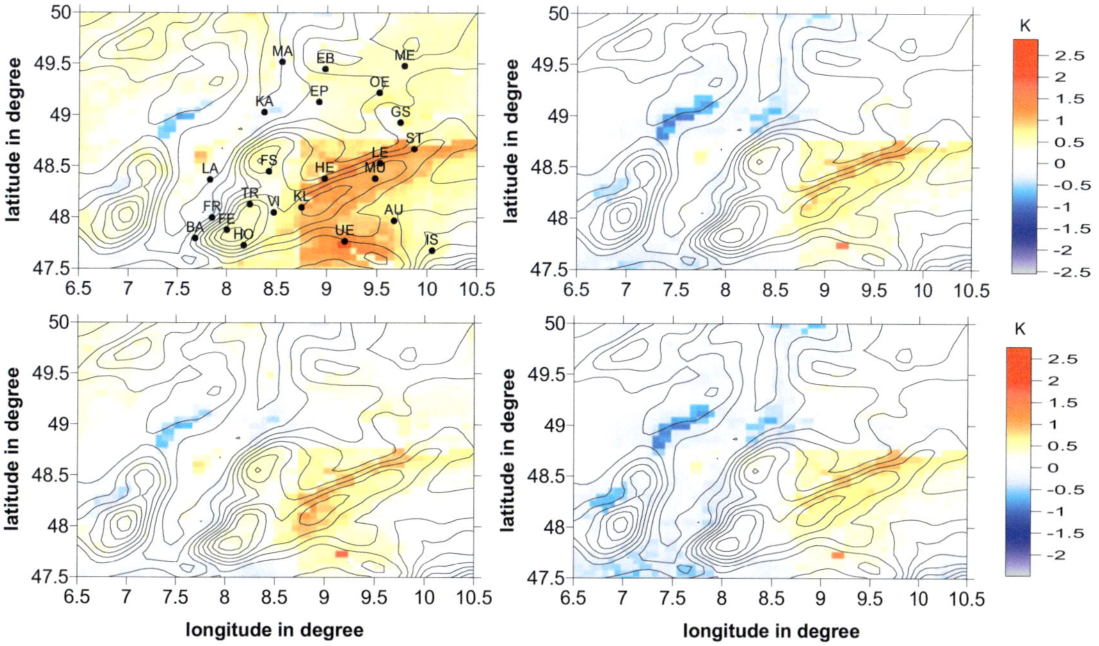

Figure 7.8: Difference of annual mean 2m-temperature between OB and MB simulation for the year 2001 for the GpSpn (upper left), Gp (upper right), GpSp (lower left) and Gpn (lower right) combination. Observation sites are indicated by points (for explanation of the abbreviations, see Tab. 7.3).

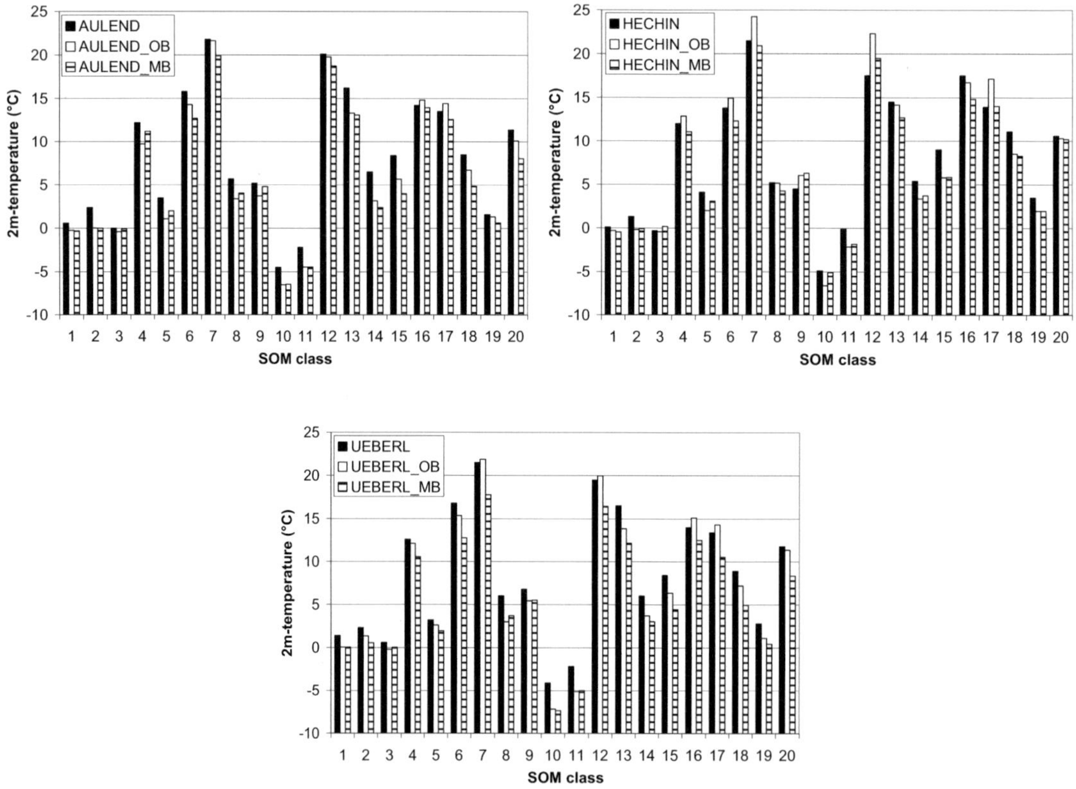

Figure 7.9: Comparison of observations (black), MB and OB simulations of 2m-temperature for the GpSpn combination for all days classified for this combination for the Aulendorf (top, left), Hechingen (top, right) and Überlingen site (bottom).

In Aulendorf we get a higher daily mean 2m-temperature and therefore better results with VEG3D initialisation for six classes compared to observations and for nine classes with ERA-40 initialisation. For five classes, the results of OB and MB simulations are nearly the same for 2m-temperature. For Hechingen, we get better results for the MB simulation for eight classes and better results for the OB simulation for nine classes and indifferent results for three classes. For Überlingen, 15 classes are simulated better in the OB simulation and only two classes are simulated better with MB initialisation.

The temperature is not underestimated for all classes of the sites in the south-eastern area so that the driving data, and therefore, the soil profiles are not inadequate for the whole year. Nevertheless, further comparisons of the NCEP driving data and the results obtained by the LSS with observed values for this region should be carried out to determine the reason for this underestimation of temperature on some days. A height correction of driving temperature should also be taken into account for the LSS simulations. When the altitude of the ERA-40 grid point is higher than the grid point of the regional model it is obvious that the pre-processed soil temperature is too low at this grid point.

Simulation	a	b	r
GpSpnOB	0.83	0.56	0.82
GpSpnMB	0.94	-0.81	0.83
GpSpOB	0.79	0.17	0.8
GpSpOB	0.89	-0.89	0.82
GpnOB	0.82	-1.2	0.78
GpnMB	0.92	-2.18	0.81
GpOB	0.81	0.28	0.79
GpMB	0.92	-0.94	0.81
ERA	0.95	-0.38	0.84

Table 7.4: Linear correlation coefficient (r), slope (a), and intercept (b) of the annual mean of 2m-temperature for statistical-dynamical simulations and the dynamical simulation.

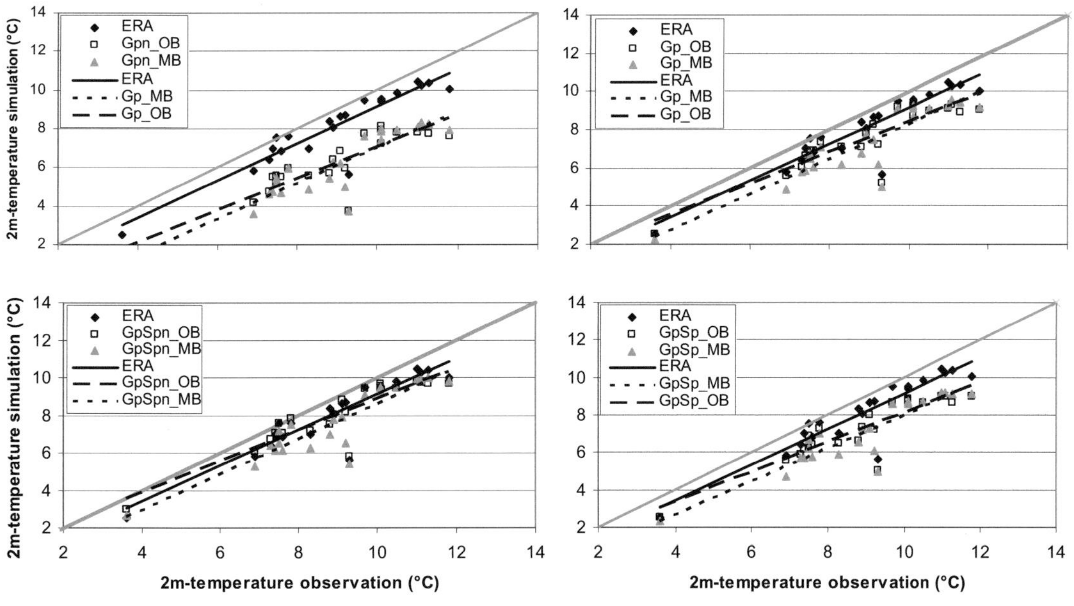

Figure 7.10: Scatter plot of the annual mean 2m-temperature of observation and dynamical (ERA) and statistical-dynamical downscaling simulations. The grey line is the bisecting line and the other lines are the linear fits.

The correlation of all four MB initialisations with the observations is better than the correlation of OB simulations (Tab. 7.4). The slope of the linear fit is closer to one and the fit is nearly parallel to the fit of the continuous ERA-40 simulation (Fig. 7.10). The initialisation by pre-processed soil moisture and soil temperature profiles therefore leads to a better correlation with observed 2m-temperature, even if the absolute temperature values are too low compared to the OB simulation, which can be seen by the intercept of the fits.

7.2.2.2 COMPARISON OF THE RESULTS OF THE TWO DOWNSCALING SCHEMES FOR ANNUAL PRECIPITATION

For precipitation, HALLER (2005) found that none of the four combinations of geopotential and surface pressure gives better results than the other. In the simulations presented here, an overestimation of averaged precipitation amount can be noticed for the dynamical downscaling, the Gpn and the Gp simulations and an underestimation for the GpSp and GpSpn simulation for OB and MB simulations (Tab. 7.5). The use of VEG3D soil moisture and soil temperature profiles leads to a better prediction of the averaged annual precipitation amount over all sites for all four combinations than the use of ERA-40 data. The differences in the mean annual precipitation amount between OB and MB simulations are in the order of 1 to 61 mm. The best results compared to observations for the mean precipitation amount is obtained by the combination GpMB. Better results are found for the MB simulations than for the OB simulations compared to observations not only for the averaged precipitation amount for this combination, but also for the majority of the precipitation amounts at the single sites.

Ordering the combinations by their accordance to the mean precipitation in contrast to the findings for the 2m-temperature, the order is not the same for OB and MB simulations. For OB simulations, the GpSp combination fits best, followed by the Gp, the Gpn and the GpSpn combinations. For MB simulations, the Gp combination fits best, followed by the GpSp, the Gpn and the GpSpn combinations. The worst result of all simulations is obtained by the dynamical downscaling, which overestimates the average precipitation amount by 335 mm. The dynamical downscaling gives higher precipitation amounts at most of the sites; for the statistical-dynamical downscaling there is an even distribution of over and underestimation at single sites. Most striking are the worse results for the GpSpn combination, which had shown best results for 2m-temperature. The strong underestimation is due to the classification of five summer days, from which four classes have precipitation amounts lower than 1 mm per day, but all together have a frequency of more than 25%. However, these summer days lead to the good performance of this combination for the 2m-temperature. This means, that for some combinations we get better results for one variable, due to the wrong representation of another variable.

Station	obs	ERA	GpSpn OB	GpSpn MB	GpSp OB	GpSp MB	Gpn OB	Gpn MB	Gp OB	Gp MB
AULEND	916.5	1332.7	742.42	814.7	1356.57	1246.38	1489.43	1544.42	1433.19	1323.19
BADENW	1240.3	2286.77	756.45	1144.32	987.84	1160.99	1498.45	1497.03	1227.88	1398.87
EBERBA	1267.2	1559.11	1135.89	1198.44	1049.39	1039.29	1036.22	1133.86	1184.16	1162.7
EPPING	924.2	1164.41	655.66	806.94	714.14	730.38	919.52	854.02	865.82	900.89
FELDBE	1944.1	2095.75	2101.22	2112.97	1993.73	1841.15	2732.35	2626.06	2368.28	2246.67
FRHBF	1127.6	1900.03	760.58	993.77	891.76	884.81	1631.95	1506.08	1079.62	1106.25
FST	1887.7	1607.7	983.62	997.75	1165.96	1127.7	1814.53	1827.62	1298.91	1279.61
GSCHWE	1263.0	1423.02	778.72	793.6	1229.39	1229.06	1001.85	953.25	1402.37	1401.31
HECHIN	916.2	1459.55	679.39	684.94	1057.49	872.16	1219.97	1349.5	1186.99	1009.43
HOEHEI	1157.6	1741.27	1535.07	1645.33	1639.77	1565.98	2073.8	2047.01	1877	1816.73
ISNY	1887.5	2719.07	908.12	934.04	1637.02	1513.65	2216.43	2243.17	1591.77	1464.43
KARLSR	873.1	1112.17	762.07	878.93	631.1	756.7	853.08	846.05	797.92	911.47
KLIPPE	1147.4	1444.46	1077.4	1186.74	1426.6	1384.94	1541.45	1542.89	1640.12	1594.3
LAHR	1030.3	1219.62	517.4	540.95	539.62	679.54	992.94	1063.37	598.05	746.04
LENNIN	1229.2	1553.14	749.44	731.92	1055.75	944.28	1025.17	1021.15	1212.1	1119.93
MANNHE	726.1	1214.08	682.83	752.65	898	829.4	811.8	888.42	1174.89	1115.05
MERGEN	938.6	1122.87	787.31	766.4	902.75	830.28	618.52	640.18	1029.25	954.59
MUENSI	1087.2	1241.22	663.25	574.8	1078.76	1050.13	997.84	972.51	1244.02	1218.64
OEHRIN	940.3	1592.27	870.31	937.75	1052.87	996.67	869.5	860.81	1161.67	1119.7
STOETT	1223.5	1603.91	704.6	723.08	1099.93	1077.54	836.17	843.48	1176.08	1166.26
TRBG	2196.0	1862.11	1310.37	1303.19	1648.21	1412.33	1947.8	1854.94	1856.32	1641.75
UEBERL	1107.3	1291.59	582.57	639.05	1216.62	1162.93	1397.95	1388.8	1325.4	1265.46
VILLIN	1122.9	1354.59	1011.22	1002.62	1333.07	1315.67	1782.1	1771.92	1538.2	1537.73
Mean	*1224.1*	*1560.9*	*902.4*	*963.7*	*1156.8*	*1115.3*	*1361.3*	*1359.8*	*1316.1*	*1282.7*

Table 7.5: Annual precipitation amount in mm for observations (obs), the dynamical downscaling simulation (ERA) and the statistical-dynamical simulations for the four combinations of geopotential and near-surface pressure with soil initialisation from ERA-40 data (OB) and from VEG3D profiles (MB) for the year 2001.

For the linear fit between observations and simulations, the correlation of the OB simulations with observations is higher than for the MB simulations, the slope is nearer to one and the intercept is lower, except for the Gpn simulation (Tab. 7.6, Fig. 7.11).

Simulation	**a**	**b**	**r**
GpSpnOB	0.55	225.0	0.36
GpSpnMB	0.50	345.2	0.29
GpSpOB	0.61	407.4	0.45
GpSpOB	0.50	492.9	0.43
GpnOB	0.95	189.5	0.50
GpnMB	1.02	112.4	0.52
GpOB	0.62	545.2	0.40
GpMB	0.53	625.9	0.38
ERA	0.69	705	0.45

Tab. 7.6: Linear correlation coefficient (r), slope (a), and intercept (b) of the statistical-dynamical simulations and the dynamical simulation.

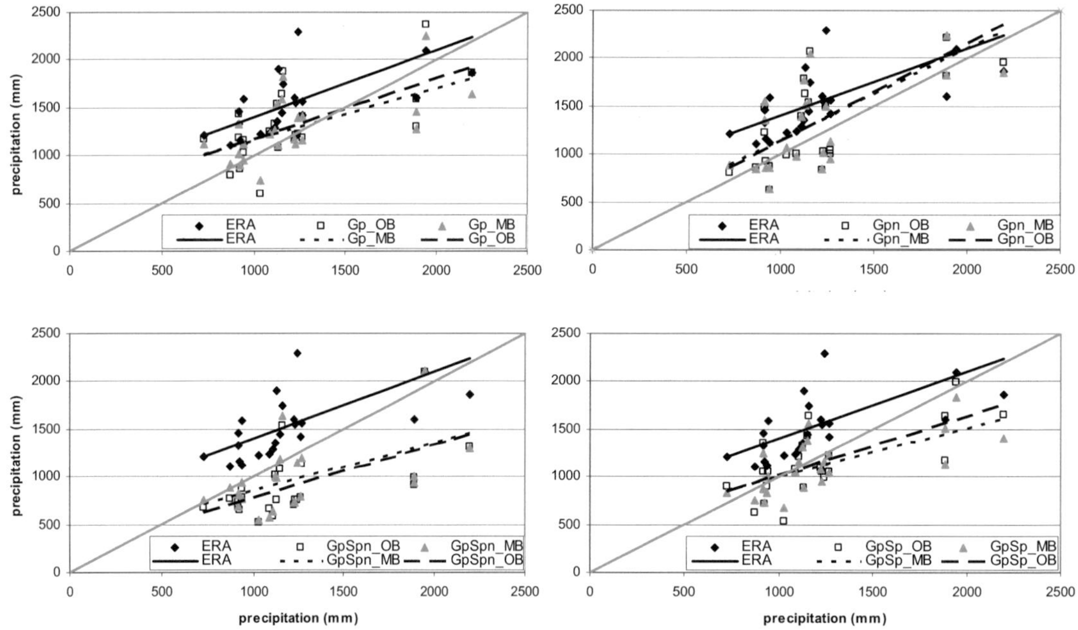

Figure 7.11: Scatter plot of the annual precipitation amount of observations (x-axis) and dynamical (ERA) and statistical-dynamical downscaling simulations (y-axis). The grey line is the bisecting line and the other lines are the linear fits.

The precipitation patterns within the investigation area are captured well by the dynamical and the statistical-dynamical downscaling compared to observations. The precipitation amounts on the ridges of the mountains are overestimated by the OB and the MB simulation for the Gp combination and the ERA simulation and the precipitation maxima over the ridges are shifted slightly to the west, which is normally the windward side of the mountains (Fig. 7.12). There is a strong overestimation in all parts of the investigation area in the dynamical simulation compared to the measurements.

The difference in the precipitation patterns for OB and MB simulations is depicted in Fig. 7.13 for the total, the convective and the grid scale amount. In general, we get an increase of total precipitation for the Gp, the GpSpn and the GpSp combinations for the west side of the Black Forest for the MB simulations compared to the OB simulations, with the strongest increase in the GpSpn simulation. A decrease in total precipitation over the ridge of the Black Forest for all four combinations with strongest differences in the Gpn simulation can be observed. For the Gpn simulation the areas with decreasing precipitation in the investigation area east of the Black Forest compensate the areas with increasing precipitation for the MB simulation in that region, so that on average we find a decrease of precipi-

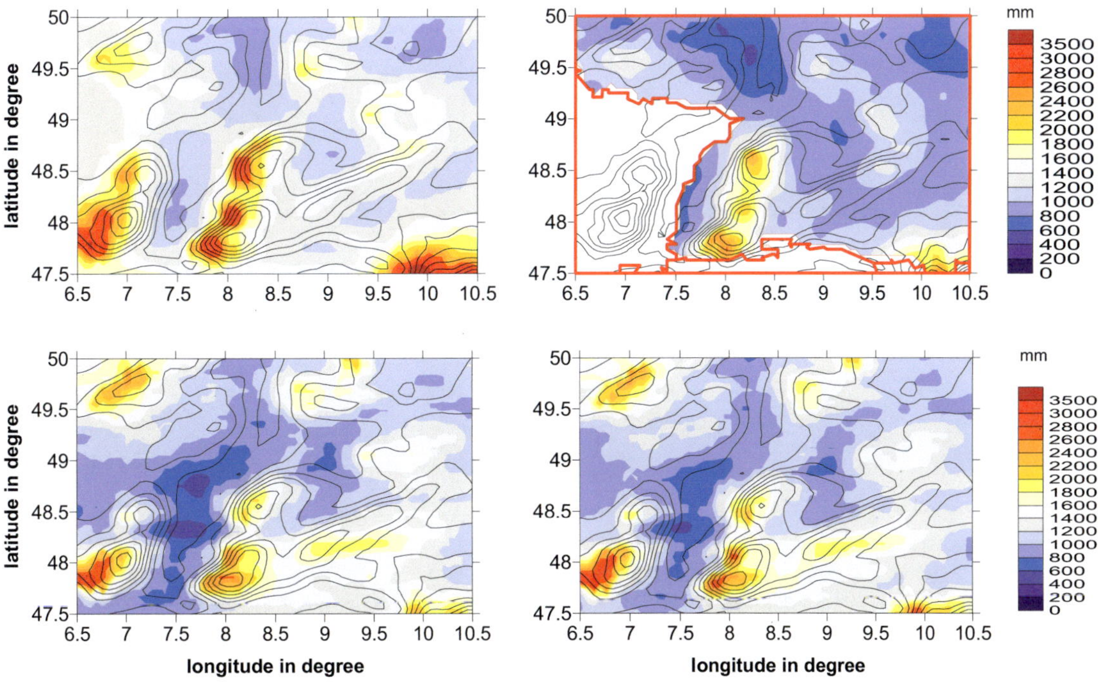

Figure 7.12: Total precipitation amount for the year 2001 for the ERA dynamical simulation (top, left), the observations (top, right), the GpOB (bottom, left) and the GpMB (bottom, right) simulation.

tation in the MB simulations there. For the GpSpn simulation we get an increase of precipitation for the MB simulations in this area instead; for the Gp and the GpSp combination the areas of increasing precipitation are as frequent as areas of decreasing precipitation in this region.

The difference in convective precipitation for all four combinations show the same patterns as the total precipitation amount, but the differences are even stronger than in the total precipitation amount for the west side of the Black Forest (Fig. 7.13).

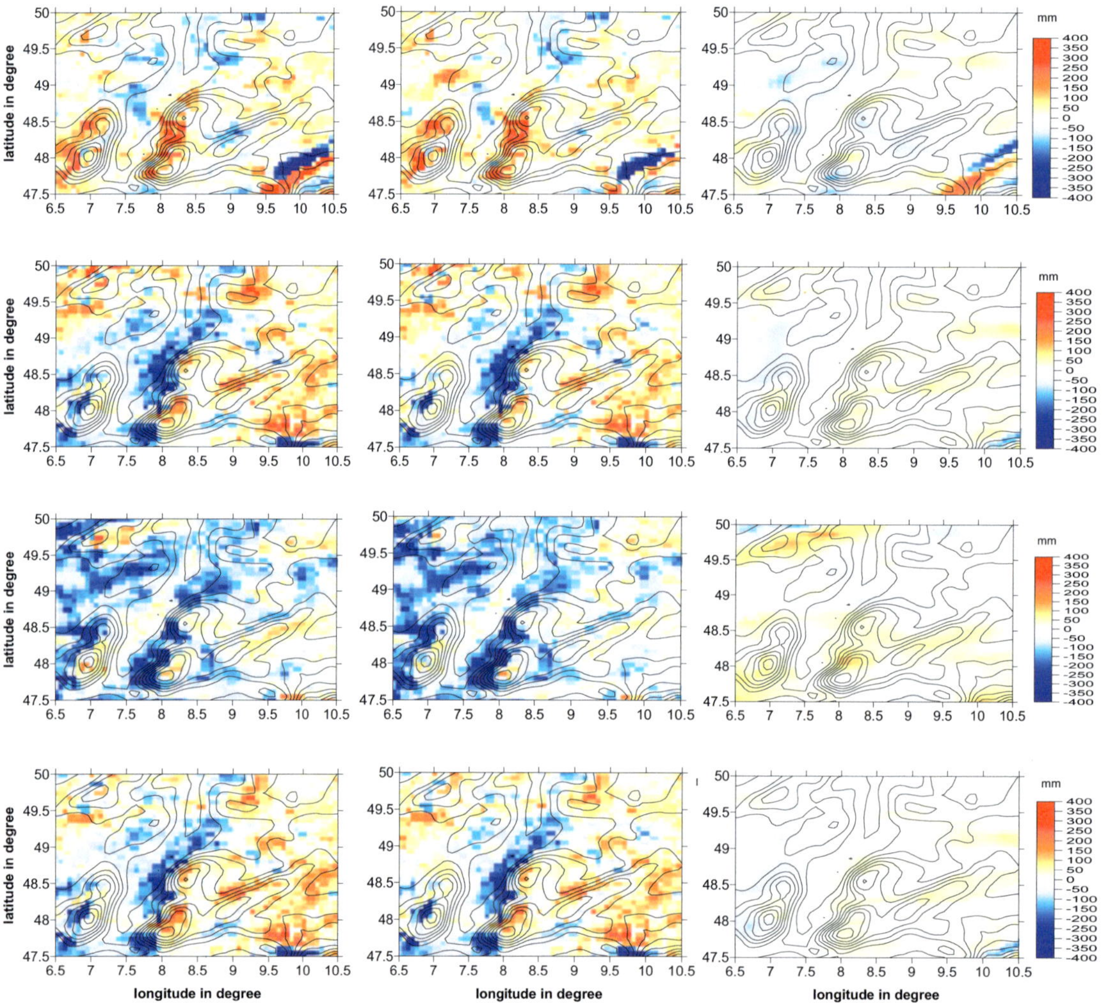

Figure 7.13: Difference of total (left), convective (middle) and grid scale (right) precipitation amount between OB and MB statistical-dynamical simulations for the four combinations Gpn (first line), Gp (second line), GpSpn (third line) and GpSp (fourth line). For abbreveations see page 137.

The differences in grid scale precipitation (max. 130 mm) are not as strong as in the convective precipitation (max. 400 mm) (Fig. 7.13). The patterns in the Rhine Valley and the Black Forest look quite similar in the Gp, GpSpn and GpSp simulations and a decrease of grid scale precipitation in the MB simulations compared to the OB simulations can be observed. The Gpn simulation is the only one that shows an increase in precipitation amount for the MB simulation over the Black Forest and the Swabian Jura.

As expected, the main differences in precipitation between the OB and the MB simulations are due to differences in the convective precipitation. To determine the influence of the soil initialisation on the precipitation amount of single classes for the Gp combination, OB and MB simulations were compared with the observed values for these classes for three sites, which are positioned in the areas with the largest differences between the two simulations.

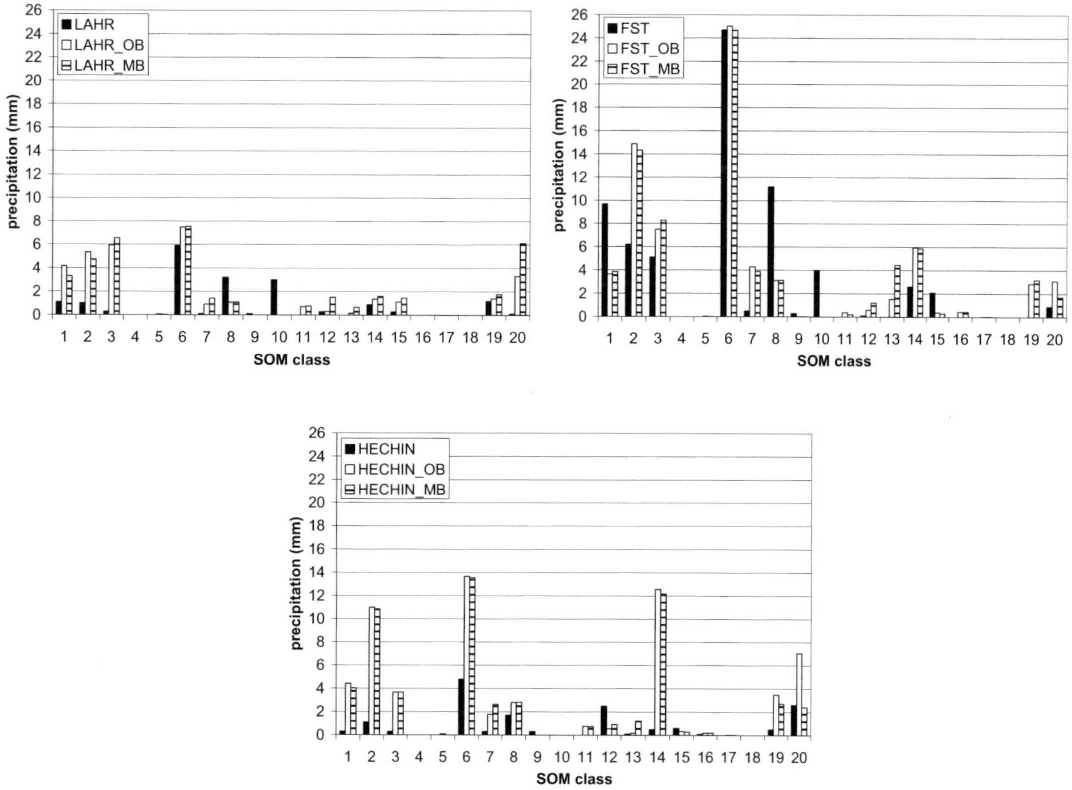

Figure 7.14: Comparison of observed (black) and simulated precipitation amount for the Gp combination for the Lahr (top, left), Freudenstadt (top, right) and Hechingen site (bottom).

For the Lahr site situated west of the Black Forest, the difference in the annual precipitation amount is about 150 mm between the two simulations with higher values for the MB simulation. This fits better with the observation but is nevertheless 280 mm too low compared to the observed value. The precipitation amounts in the simulations that are too low are mainly caused by two classes, for which the model produces less or even no rain at all, but where precipitation was observed (Fig. 7.14). The higher precipitation amounts of the simulations obtained in other classes cannot compensate for this difference due to their lower frequency. The precipitation amounts for the single classes are quite similar in both simulations for most of the classes and we get maximum differences of 2.5 mm. In eight of the 13 classes, for which precipitation is simulated, precipitation amounts are enhanced in the MB simulation. This enhancement is unrealistic compared to the observations.

For the Freudenstadt site, which is situated in the Black Forest, the difference in annual precipitation amount is about 20 mm between the OB and MB simulations with better results for the OB simulation. The simulated precipitation amount is about 600 mm lower than the observed one. The number of classes (four) where the precipitation amount fits better to the observed one in the OB simulation is equal to the number where it fits better for the MB simulation (Fig. 7.14). As for the Lahr site, we get lower and no precipitation in the simulation for class 8 and 10, although precipitation had been observed. This leads to the underestimation of the annual precipitation amount by the simulations. This cannot even be compensated by the 10 classes where the precipitation amount is overestimated by the simulations due to the high frequency of class 8 and 10.

For the Hechingen site in the east of the Black Forest, there is an overestimation of the annual precipitation amount by the simulations of about 100 mm. This is due to an overestimation of precipitation by the simulations in nearly all classes where precipitation is observed (Fig. 7.14). The results for the OB and the MB simulations differ by maximum 1 mm in all classes except class 20. In class 20, the difference is about 5 mm with unrealistic precipitation amounts for the OB simulation and quite good results for the MB simulation. As this is the most frequent class, we get a difference of 180 mm between OB and MB simulation and better results compared to observations for the MB simulation. For seven classes the MB simulation gives better simulation results than the OB simulation compared to observations and the OB simulation only shows better results for one class.

From these three examples it may be concluded that adequate annual precipitation sums are sometimes obtained for the wrong reasons. The model is not able to simulate the precipitation amounts correctly for each class but the underestimation of precipitation amount by some classes compensates the overestimation by other classes. The difference in precipitation amount is lower than 1 mm for most of the classes between OB and MB simulations. The influence of soil moisture and soil temperature on precipitation is largest in summer for days in May and June (class: 13 and 20). On these days, the differences in precipitation amounts between OB and MB simulations are strongest. For some sites and classes the use of VEG3D soil profiles leads to an enhancement of precipitation and sometimes to a decrease. This fits better to observations for some site/class combinations and not for others. Due to the limited number of summer days, it is not possible to decide whether the method of using pre-processed soil profiles or the method of using ERA-40 reanalysis data as initialisation for the soil profiles gives better results. A statistical evaluation of daily simulations initialised with ERA-40 data and initialised with VEG3D data for summer days of different synoptic conditions is necessary to judge the efficiency of both methods. For non convective days, the initialisation by VEG3D soil variable profiles seems to work well when only precipitation is compared.

7.3 SUMMARY

In this chapter it has been shown that the statistical-dynamical downscaling method presented here is able to reproduce the results for 2m-temperature obtained by dynamical downscaling and it shows even better results for the precipitation amount than the dynamical downscaling. This may be caused by the development of an own climate in the dynamical simulations where feedbacks with evaporation over longer time scales than a single day can cause higher precipitation rates on some days. The underestimation of 2m-temperature at all sites with the statistical-dynamical downscaling compared to observations is mainly due to the too low temperatures in the driving data for the daily simulations. The average precipitation amount is reproduced reasonably well by statistical-dynamical downscaling, although differences of up to 700 mm in the annual precipitation amount can occur at the individual sites.

The GpSpn combination generates the most reliable results for 2m-temperature as does the Gp combination for precipitation. This difference in suitability of one combination for 2m-temperature and precipitation simulation is mainly due to the over or underrepresentation of summer days by the different combinations. Due to this sensitivity to classified days, it is not possible to decide in advance for any other year, which of the four combinations gives best results and, if a combination can be found which fits best for 2m-temperature and precipitation amount. Ideally, for a practicable statistical-dynamical downscaling method with focus on temperature and precipitation, a SOM classification should be developed, which guarantees best fits for temperature and precipitation so that only simulations for one combination are necessary. This could be achieved by using more than two meteorological fields for the classification of the 20 classes or by taking into account more than 20 classes to better represent summer days with convective precipitation. It would be also possible to force the algorithm to identify important classes or to determine the number of classes, which are to be classified in a certain period of the year. Further meteorological fields for classification could be the relative humiditiy in 700 hPa or the equivalent-potential temperature between 850 and 500 hPa (HEIMANN and SEPT, 2000). An increase in the number of classes would prevent a too strong influence of single classes on the result and perhaps offer the possibility to take into account more convective summer days. Increasing the number of classes from 20 to 30 would mean an increase of 50% in computing time, but this is only an increase from 5% to 8% in the computing time needed for a dynamical simulation. Even if different combinations for the calculation of the 2m-temperature and for the calculation of the precipitation amount would be necessary, the method would save 80 % of computing time compared to the dynamical method.

The use of pre-processed soil moisture and soil temperature profiles seems to be an advantage. However, the values of the soil variables depend largely on the quality of the driving data of the stand-alone model. Too coarse driving data may be a disadvantage because it affects a considerable part of the investigation area and if the data are not consistent, the results of the statistical-dynamical downscaling in this part may be worse than with soil profiles taken directly from the driving data. Possibilities to overcome this problem would be the use of driving data with a finer horizontal resolution or the height correction of driving data at least for temperature. Further steps in the development of a more homogeneous statistical-dynamical downscaling method could be the use of the same data for SOM classification and boundary data for the driving model and the use of the same LSS for stand-alone and regional model simulations. In this study the SOM classification has been done using the NCEP reanalysis data and ERA-40 reanalysis data has been used as driving data for the regional model. This could be harmonised by using the same reanalysis data as data for classification and as driving data. The LSS VEG3D was used for the production of the adjusted soil profiles; the COSMO-CLM uses TERRA_LM as LSS. In principle, it is also possible to use TERRA_LM for the production of the adjusted soil profiles.

8 SUMMARY AND OUTLOOK

Until today, very few high-resolution regional climate simulations with horizontal resolutions finer than 15 km have been carried out and many questions about the advantages and uncertainties of such simulations regarding meteorological as well as hydrological quantities still remain open. In this thesis, basic investigations on high-resolution regional climate modelling over complex terrain were carried out to assess the advantages and uncertainties of high-resolution regional climate simulations; this was done by performing ensemble simulations. The focus is on the sensitivity and reliability of the 2m-temperature as an important quantity to detect climate trends and on precipitation as an important quantity of the hydrological cycle. Both quantities may be compared to observations and are therefore suitable to evaluate the high-resolution simulations and to determine the added value compared to simulations with coarser horizontal resolution.

Three main topics were investigated, with the aim of promoting high-resolution regional climate simulation with the regional climate model COSMO-CLM. First, an adequate model setup for high-resolution regional climate simulations for the area of Southwest Germany was defined by the evaluation of ensemble simulations with the regional climate model COSMO-CLM. Secondly, an advanced land surface scheme with an explicit vegetation layer was coupled to COSMO-CLM to improve the parameterisation of the soil and vegetation processes, and to study the influence of land surface schemes on regional climate simulations and particularly on the water cycle. Third, a statistical-dynamical downscaling method was developed to replace the time-consuming dynamical simulations.

The regional climate model COSMO-CLM was used for all the studies. The model has been developed in recent years from the weather prediction model COSMO of the German Weather Service. So far, the model has never been used with horizontal resolution finer than 10 km for climate applications over decadal periods for Southwest Germany. The 1990s were selected as an evaluation period since the climate change is most pronounced in recent years and the availability and quality of driving data and observations is highest during this period. The necessity and strengths of high-resolution climate simulations can be demonstrated best for areas of complex topography. Therefore, the evaluation has been done for the region of Southwest Germany.

A model setup for climate simulations can be considered suitable when the simulation yields best results compared to the observations of past decades. Based on the performance of ensemble simulations over the period from 1988 to 2001 and by varying domain size, horizontal resolution, driving data, physical parameterisations, and time integration schemes, an adequate model setup was determined. A suitable model setup includes the following: ERA-40 reanalysis data as driving data, a horizontal resolution of at least 7 km, a model domain containing the Alps, and climatological soil water and soil temperature initialisation. The better performance of higher resolution simulations compared to simulations with coarser resolution has been clearly shown in this study. The Runge-Kutta time integration scheme can be considered as a promising alternative to the leapfrog scheme but further studies should be carried out to demonstrate the benefit of this scheme, which was originally used only for horizontal resolutions finer than three kilometer.

The variability obtained by using different model setups is largest for the change of the driving data at the lateral boundaries of the model. For temperature, the regional model is able to correct differences between the driving data due to the production of an own climate, which is more influenced by the processes within the simulation domain than by the data given at its lateral boundaries. For large scale phenomena like cyclones, the differences in the driving data sets cannot be compensated by the regional model and are therefore reflected and even amplified in the precipitation results of the regional model. Hence, differences between simulations with different driving data sets are largest in winter time when the large-scale forcing dominates the circulation in the regional climate model. In summer, when the large-scale forcing is weak, the regional model results are much more independent of the driving data set. Overall, the ERA-40 reanalysis data set provides better driving data for the selected region than the NCEP reanalysis data set.

The variability obtained by changing the initial soil moisture content, the number of active soil layers and greenhouse gas concentration are quite small and a clear sensitivity is only observed for the change of the convection scheme and the time-integration scheme.

The influence of simulation setup on single components of the water balance of the soil is quite different; evaporation and soil water content are rather unaffected by the change of horizontal resolution or driving data, whereas precipitation and runoff show a clear dependency on such changes.

All ensemble simulations show a negative bias in 2m-temperature and a positive bias in precipitation, which has also been detected by other model studies with COSMO-CLM (BÖHM et al., 2006). The bias in temperature is about -0.8 K and in precipitation about 20%. For horizontal resolution of 50 km, model intercomparison studies found a bias of +/- 2°C and a precipitation bias of +/- 50% for regional climate models (IPCC, 2001). Comparing the bias obtained in COSMO-CLM simulations with these findings, the results obtained by high-resolution climate simulations with COSMO-CLM are more than satisfying. The still significant overestimation of precipitation compared to observations, especially in winter, may also be a problem of the observation error that leads to an underestimation of observed precipitation in winter.

With the adequate model setup, COSMO-CLM is able to reproduce the orders of magnitude of the observed temperature trends of the last decade.

In this thesis, the operationally used land surface scheme TERRA_LM was replaced by the VEG3D land surface scheme in COSMO-CLM. To provide a VEG3D version that is adequate for climate simulation, the freezing and melting processes in the soil were implemented in VEG3D. Freezing and melting processes in the soil are important for the correct simulation of soil temperature and soil water content and therefore for the calculation of the heat fluxes into the atmosphere. Stand-alone simulations with the newly implemented parameterisation show good agreement to observations and better results than the version without soil freezing processes.

In most models, within the soil column of a regional model grid box only one soil type is considered for the whole column, although the soil generally consists of several layers with different soil types. The use of different soil types within one soil column improves the simulation results of the stand-alone VEG3D version with respect to the soil water transport considerably. Therefore, a soil-type inventory was provided in this thesis, which can be used for simulations with the COSMO-CLM/VEG3D system.

Better agreement of soil temperature and heat fluxes with observations is obtained in stand-alone simulations with the VEG3D scheme compared to the TERRA_LM scheme, especially over high vegetation. This is due to the explicit consideration of a vegetation layer and a different formulation of soil water transport and root depth in the VEG3D scheme.

The performance of the COSMO-CLM/VEG3D system was assessed by single-day simulations. The near-surface parameters, such as e.g. heat fluxes and 2m-temperature, show reasonable results and also the evolution of the boundary layer that depends on the heat fluxes, calculated by the land surface scheme, seems to be realistic.

The findings for the annual simulation with the online coupled system, which were driven by GME analysis data, are similar to the findings for the decadal simulation, which were driven by ERA-40 reanalysis data. The area average of annual precipitation shows lower amounts in the VEG3D simulation, which is in better agreement with observations. The convective precipitation increases in the simulation with VEG3D compared to the simulations with TERRA_LM and the grid scale precipitation decreases. Due to the nonlinearity of the processes involved, several reasons are responsible for this behaviour. One possible reason is the reduced evaporation over the land areas, which leads to a lower humidity in the atmosphere, and, therefore, to a reduction of grid scale precipitation in the simulation with VEG3D. For the triggering of convection, and therefore also for the production of convective precipitation, the heating of the soil is important. The heating is higher in the VEG3D simulation in the Rhine Valley and the slopes of the Black Forest. This may lead to more frequent convective precipitation and to an increase in the convective precipitation amount. The differences in annual 2m-temperature and precipitation in VEG3D and TERRA simulation are comparable to the differences between the TERRA simulation and simulations, where the convective scheme or the time-integration scheme are changed. Neither of the two land surface schemes gives clearly better results for 2m-temperature and precipitation than the other in the online coupled simulations. Due to the fact that observations of heat fluxes, soil temperature and soil water content are quite rare, the model performance in regard to observations can only be investigated by comparing 2m-temperature and precipitation. In the same vein, the differences in the water balance can only be investigated qualitatively: the area averages of the absolute values in both simulations are quite similar with higher precipitation and evaporation amounts in the TERRA simulation (difference of 3 and 8% in the decadal sum) and higher runoff rates in the TERRA simulation (difference of 3% in the decadal sum). The increase in soil water content over the decade is higher in the VEG3D simulation; that is due to the higher water holding capacity of the soil. Although the differences in the area averages are below 10%, larger differences in the water balance components for single parts of the model area are found. Therefore, the use of a different land surface scheme yields a noticeable effect on the water cycle and this influence of land surface scheme on hydrological variables should be kept in mind when studying the water cycle with meteorological models.

In this thesis, a strategy for an efficient statistical–dynamical downscaling scheme was developed for regional climate simulations. The SOM algorithm is used to select single days within a period, which are representative for the circulation patterns occurring during this period. These days, classified by the SOM algorithm, are simulated explicitly by the COSMO-CLM and the results are added up, weighted with the frequency of the pattern represented. This can save about 80% of the CPU time needed for an explicit simulation of every day of the year; however, up to now only averages over the time period can be obtained this way. The problem of the soil initialisation was solved by using pre-processed soil moisture and soil temperature profiles. These profiles are produced by running the soil model in stand-alone mode driven by the reanalysis data used also as driving data for the daily simulations. These simulations are initialised some years before the starting date of the daily simulations and the soil profiles can therefore adapt to the land surface scheme and the driving climate.

The statistical-dynamical method provided here is able to reproduce the results obtained by dynamical downscaling for the annual means of 2m-temperature and precipitation. Compared to observations, both methods underestimate the 2m-temperature and overestimate the precipitation. This disagreement of statistical-dynamical downscaling with observations is not caused by the classification of days, however. This has been shown by using the observed daily mean values for the calculation of the annual means by the statistical-dynamical method. With observed values, the results of the statistical-dynamical downscaling agree well with the annual mean values obtained from continuous observations. The disagreement is rather caused by the driving data as demonstrated by the performance of simulations with different driving data sets.

The use of pre-processed soil temperature and soil water profiles shows the potential to improve the single day simulations but the profiles are strongly influenced by the driving data especially during the summer months and, therefore, further studies are necessary to investigate and improve the quality of the driving data.

In this thesis, a first step towards reliable high-resolution climate simulations with COSMO-CLM was made, which can serve as the basis for further studies. Simulations using the adequate model setup, but with a horizontal resolution of 2.8 km, and ensemble simulations, where the initial and boundary conditions will be disturbed, are planned at the institute to study present and future climate change on regional scales. The Runge-Kutta scheme seems to be an alternative to the leapfrog scheme and further studies should be carried out to confirm the advantages and disadvantages of this time-integration scheme.

Further evaluation of the VEG3D coupled online to the COSMO-CLM is planned and a reconsideration of the calculation of near-surface parameters will be necessary to provide near-surface values more suitable to the observed ones over high vegetation. A reconsideration of the boundary layer parameterisation may be useful to take better account of the additional vegetation layer. Until now, three-dimensional maps were provided to take into account different soil types within one soil column, but the evaluation of coupled simulations using these maps needs to be undertaken.

Statistical-dynamical downscaling reveals a high potential to replace time-consuming dynamical simulations for the provision of annual mean values of meteorological variables, but the classification of single days has to be reconsidered and further studies should be carried out on the initialisation of the soil profiles.

9 APPENDIX

9.1 ASSIGNMENT OF SOIL TYPES FROM HAD TO VEG3D

In the Hydrologischer Atlas Deutschland (HAD, 2003) 60 soil forms are classified for the area of Germany depending on their geological history (Fig. 8.1). Each soil form consists of at least one soil horizon and each soil horizon consists of one or more different soil types (Fig 8.2). To use the information from HAD in VEG3D, the 34 soil types defined in HAD for each soil form are assigned to the 13 soil types used in VEG3D (Tab. 8.1). A digitized map of HAD is then used to produce a three dimensional digital inventory of VEG3D soil types. The soil classification of HAD ranges down to a depth of 2 m. Below this depth, loam is assumed as soil type for all VEG3D layers.

soil type VEG3D	soil type HAD
loamy sand (lS)	Sl4, Sl3, lu, Sl2, St2, Su4, Su3, Su2
loam (L)	Ls2
silty clay loam (utL)	Lt3
clay loam (tL)	Lt2
clay (T)	Tl, Tt
peat (Tf)	Hn, Hh
sandy loam (sL)	St3, Ls4, Ls3
sand (S)	Ss, fS, mS, gS
silt loam (uL)	Uls,Lu, Ut4, Ut3
sandy clay loam (stL)	Lts
silt (U)	Ut2
sandy clay (sT)	Ts4, Ts3, Ts2
silty clay (uT)	Tu4, Tu3, Tu3

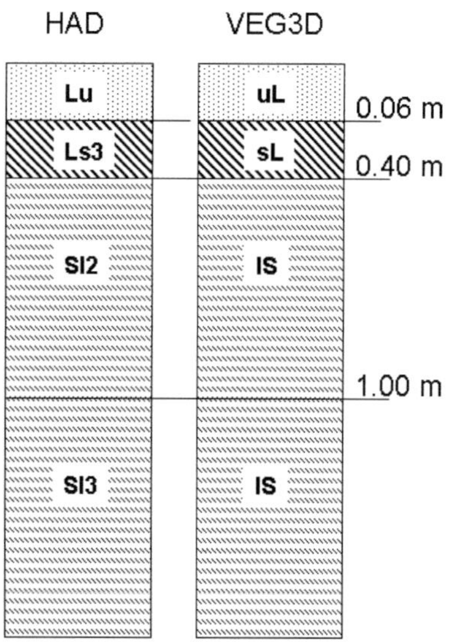

Table 9.1: Assignment of HAD soil types to VEG3D soil types (left) and example for soil type assignment for soil form 13 (right).

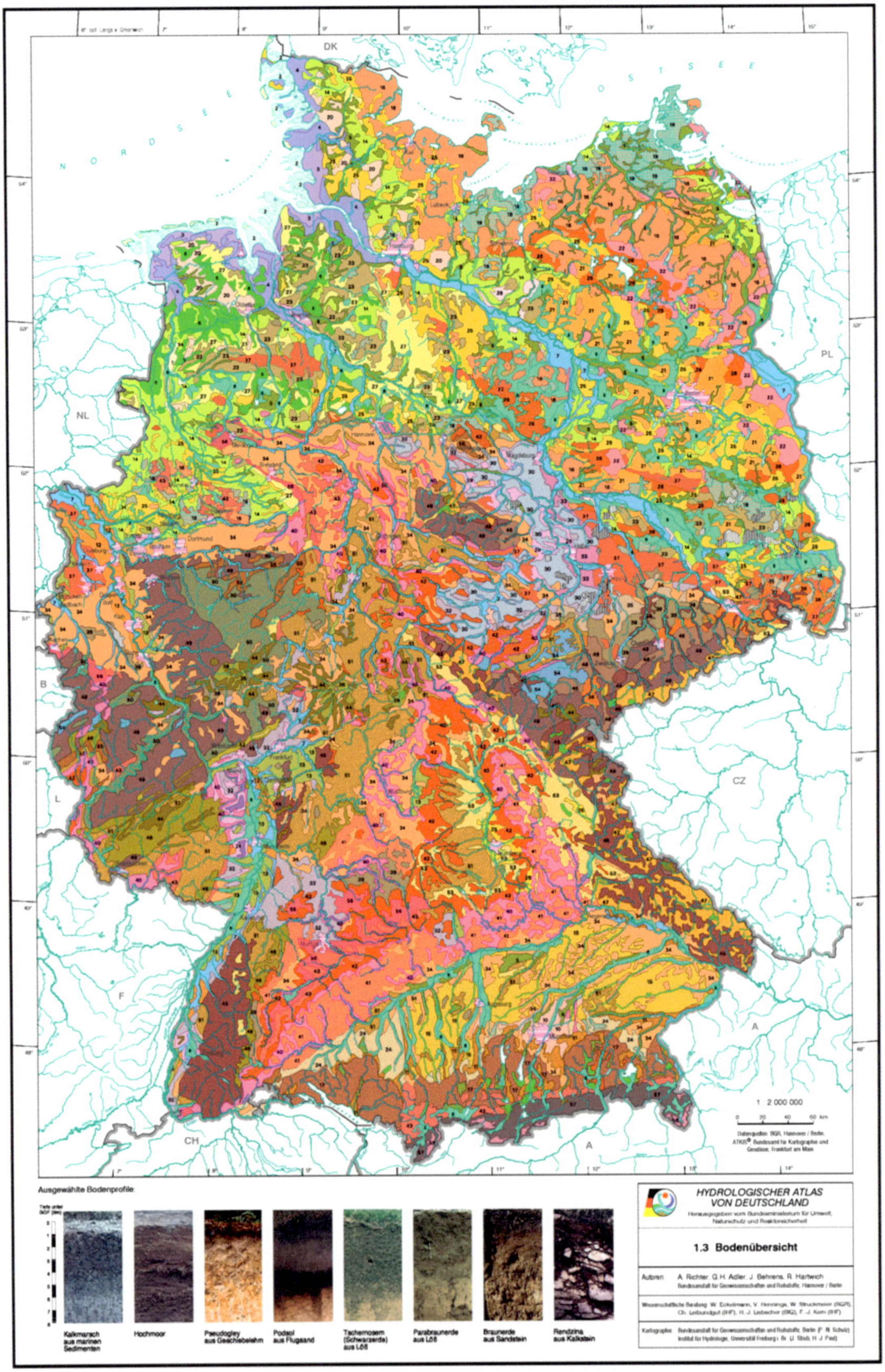

Figure 9.1: Map of soil forms in Germany from HAD (Courtesy of RICHTER et al., 2003).

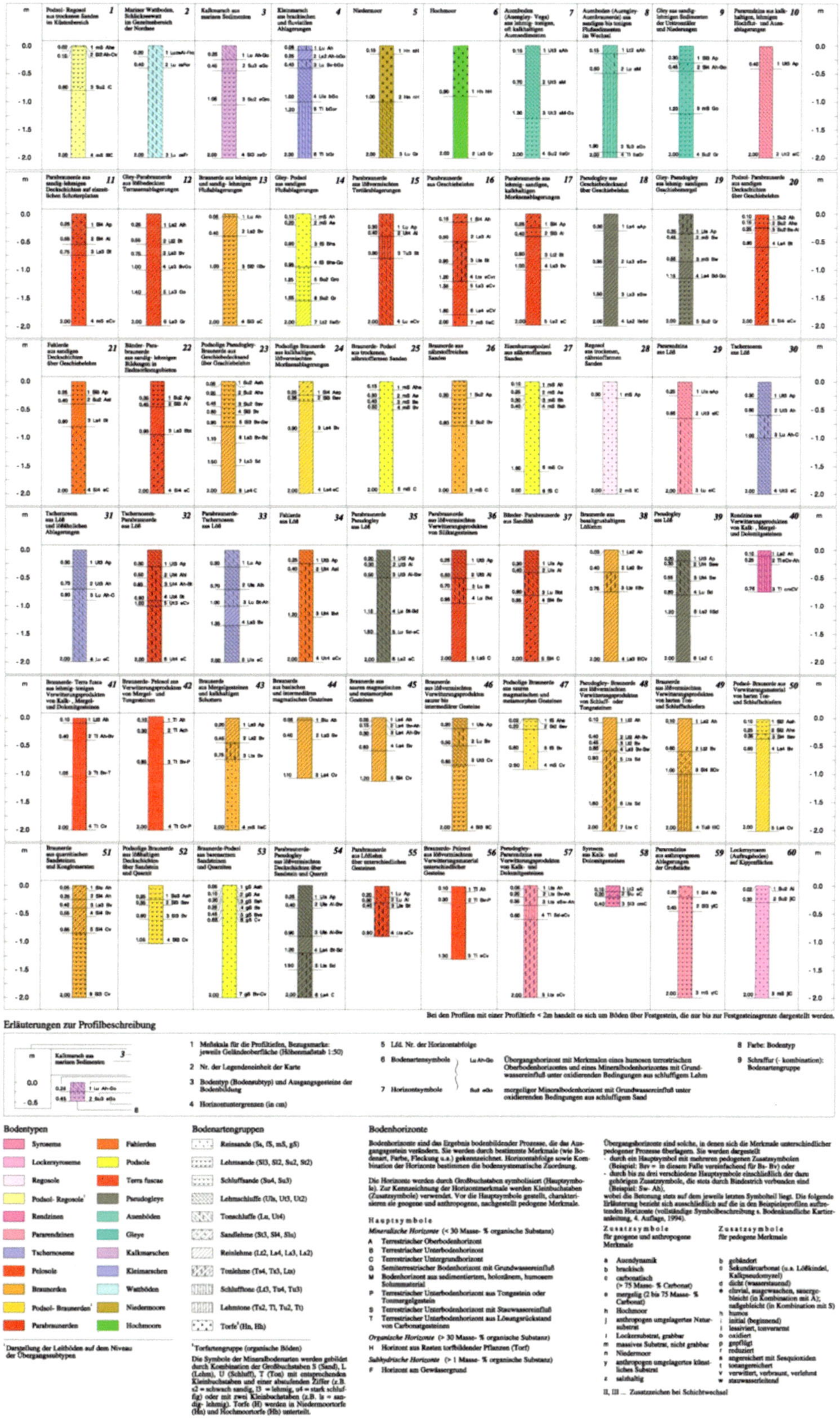

Figure 9.2: Soil forms with assigned soil types from HAD (courtesy of RICHTER et al., 2002).

9.2 EVALUATION OF THE REVISED CLOUD MICROPHYSICS SCHEME OF COSMO MODEL

In COSMO model version 3.22 a revised cloud microphysics scheme is implemented to overcome the problem of overestimated precipitation in winter, too frequent very light precipitation, and overestimation of orographic precipitation. Main changes in the cloud microphysics scheme were:

(1) The replacement of the Kessler-type autoconversion/accretion scheme by the parameterisation of SEIFERT and BEHENG (2001)

(2) The introduction of a new parameterisation for the intercept parameter in the snow size distribution

(3) The introduction of a temperature dependent sticking efficiency and

(4) The change in geometry of snow and terminal fall velocity

These changes lead to a slower formation of rain and snow and a reduced sedimentation velocity for snow. A more detailed description of the changes made and their influence on COSMO model results can be found in SEIFERT (2007).

To evaluate the long term behaviour of the scheme, two different kinds of simulations were performed:

1) Simulations driven by GME analysis data every six hours for the period from 1.12.2000 to 31.12.2001 with a grid size resolution of 0.0625° and 64x64 grid points for Southwest Germany

2) Simulations driven by reanalysis data every six hours for the period from 1988 to 2001 with a grid size resolution of 0.44° for the whole European region (80x61 grid points)

The simulations were run with the COSMO model version 3.21 containing the old scheme and the COSMO model version 3.22, where the new scheme is implemented. The only difference between the two model versions is the change in the cloud microphysics scheme.

a) Evaluation of the GME driven simulations for the year 2001

The comparison of the total precipitation shows two main differences between the old and the new scheme (Fig. 9.3): The precipitation is transported more to the lee side of the mountains with the new scheme, resulting in a decrease of precipitation amount at the windward side of the mountains and an increase on the lee side, which is much more realistic compared to observations. The area average of precipitation increases by about 1 mm with the new scheme.

Comparisons of the grid scale precipitation between the two schemes show a decrease of precipitation amount over the Black Forest and an increase of precipitation on the lee side of the mountains in the simulation with version 3.22 (Fig. 9.4).

Comparisons of the convective precipitation show no significant differences between the two schemes (Fig. 9.5).

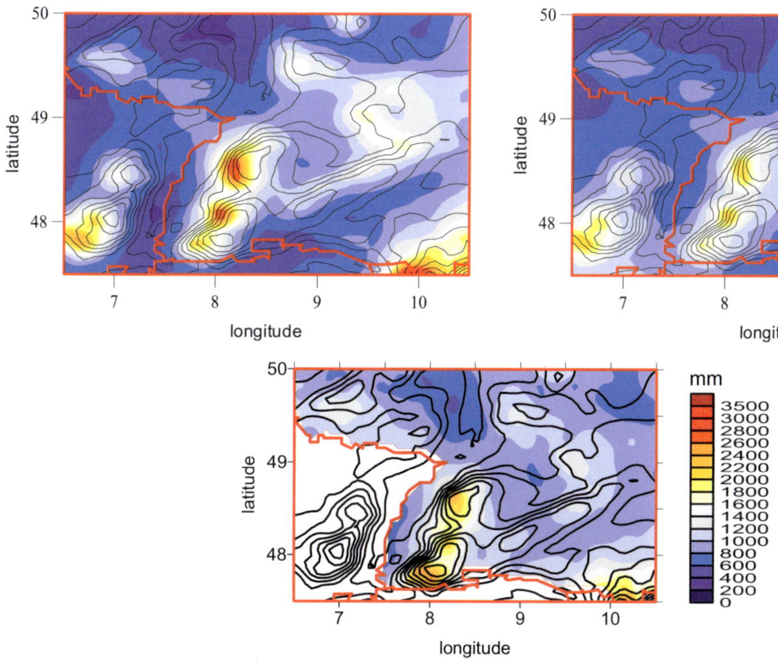

Figure 9.3: Annual total precipitation amount for the year 2001 for the old (upper left) and new (upper right) microphysics scheme for the whole area and for observations within Germany (bottom).

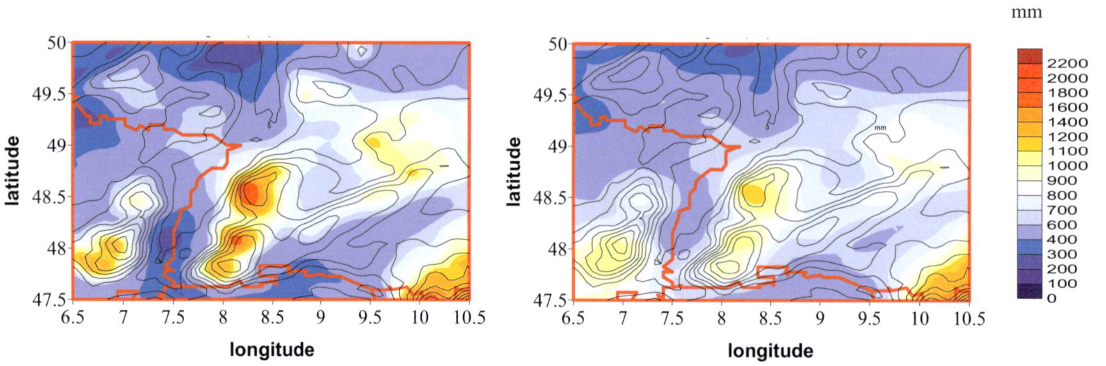

Figure 9.4: Annual grid scale precipitation amount for the year 2001 for the old (left) and new (right) microphysics scheme.

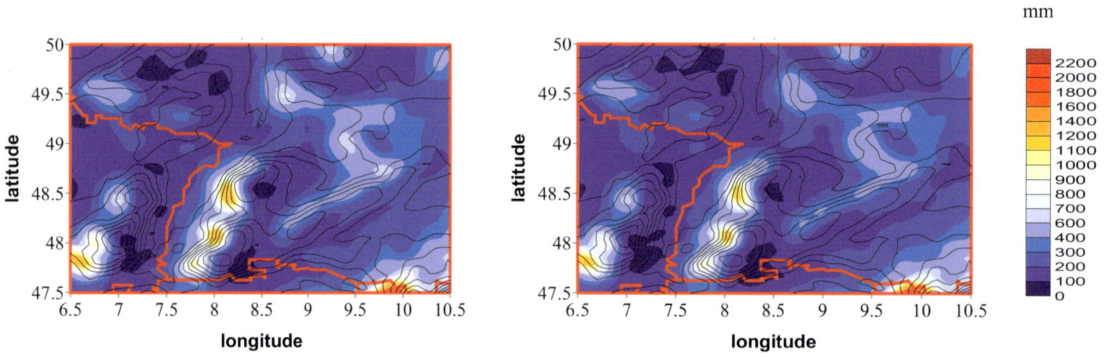

Figure 9.5: Annual convective precipitation amount for the year 2001 for the old (left) and new (right) microphysics scheme.

A comparison of the monthly total and grid scale precipitation sums shows that the largest differences between the two schemes occur in the winter months with maximum differences of about 25 mm, but there is no version, which generally produces more or less precipitation over all months (Fig. 9.6).

The convective precipitation shows a slight decrease of up to 5 mm per month with the new cloud microphysics scheme and the maximum differences occur in summer (Fig. 9.7). In the monthly mean 2m-temperature there are only marginal differences of up to 0.2 K (Fig. 9.7)

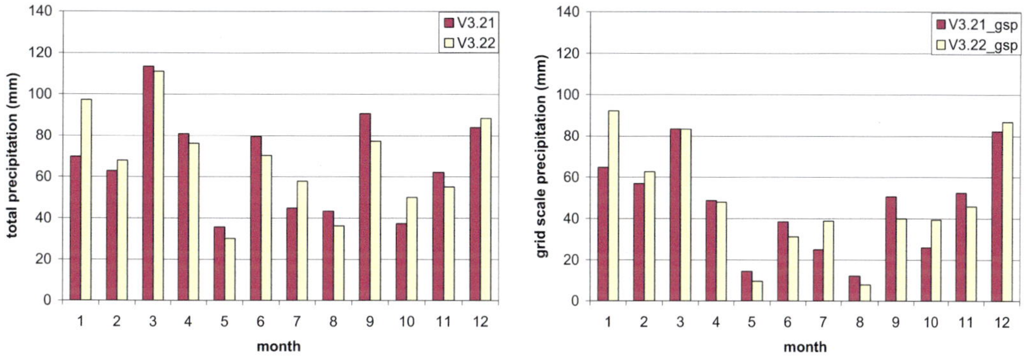

Figure 9.6: Monthly total (left) and grid scale (right) precipitation amount for the year 2001 with the model version 3.21 (red) and 3.22 (yellow).

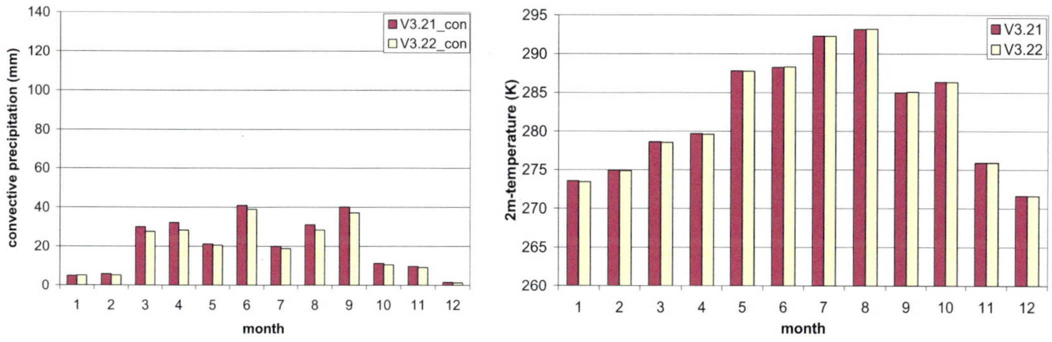

Figure 9.7: Monthly convective precipitation amount (left) and monthly mean 2m-temperature (right) for the year 2001 with the model version 3.21 (red) and 3.22 (yellow).

b) Comparison of the ERA 40 driven simulations for the period 1990 - 1999

Simulations have been done with ERA-40 and NCEP Reanalysis data as driving data and with climatological soil initialisation as described in section 4.1. For the evaluation the average of the annual precipitation sums for the period 1990 – 1999 has been compared. An increase of precipitation with the new version 3.22 can be observed for Germany, the north of France and Eastern Europe by up to 100 mm per year in the simulation with NCEP driving data (Fig. 9.8). A decrease of precipitation amount with the new scheme can be observed for the alpine region and Spain by up to 200 mm. The same findings are also valid for the simulation with ERA-40 as driving data (Fig. 9.8).

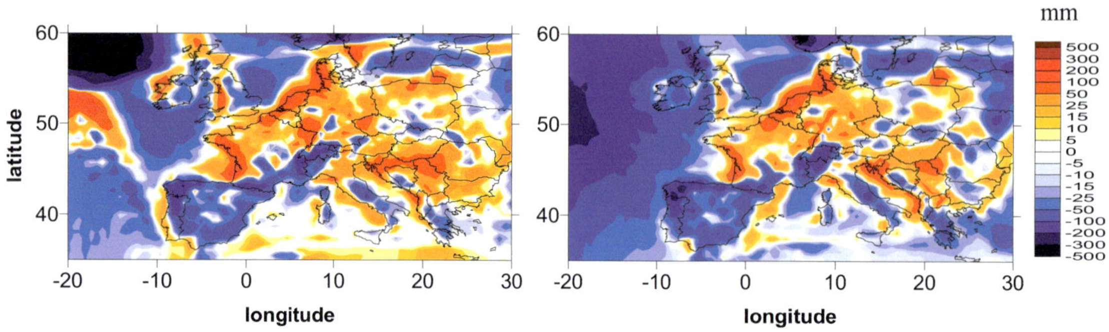

Figure 9.8: Difference of averaged annual precipitation sum for the period 1990–1999 between the new and the old microphysics scheme for simulation driven by NCEP (left) and ERA-40 (right) reanalysis data.

9.3 Influence of GME driving data on daily simulations

Since the year 2005, the GME uses the multi layer soil model and data for a correct initialisation of the multi layer soil model in COSMO model is available. Using the interpolation program int2lm it is also possible to interpolate the GME data before 2005, which contain the older two layer soil scheme (TERRA), onto the multi layer grid of the multi-layer soil scheme (TERRA_LM) in the COSMO model. Due to the formulation of the soil schemes the uppermost soil layer in the older two layer scheme contains in general more soil water than the equivalent layers in the new multi-layer scheme would do for the same simulation period. For climate simulation this is not as important because the surplus in water content is evaporated after some days, but for simulations of single days this can be of crucial importance.

Two different simulations have been performed to investigate the influence of the soil moisture initialisation onto COSMO model simulations for a single day: In the first one the multi layer soil model in the COSMO model was initialised by the data from the two layer soil model from the GME data. In the second one the soil initialisation was taken over from newly processed GME data, which already contain the multi layer soil model. In the second simulation only the soil initialisation has been taken over from the new GME data. The atmospheric forcing is the same as in the first simulation.

Simulations for the first of June 2002 with a horizontal resolution of 7 km for the region of Southwest Germany show that the results in near-surface specific humidity can differ by up to 3 g kg^{-1} between the two simulations with higher values in the simulation where the two soil layer GME data has been interpolated onto the multi-layer COSMO model. These values are too high compared to observations and the simulation where multi layer soil model data from GME was used for the initialisation of COSMO model agree better with the observed near-surface values (Fig. 9.9).

Differences are also visible in the boundary layer where we find unrealistic high specific humidity in the first kilometre above the surface. The specific humidity is up to 1 g kg^{-1} higher than in the simulation where we initialised with the multi layer data and 1.5 g kg^{-1} higher than the observed one. As a consequence, we observe an unrealistic low potential temperature, which is up to 1 K lower than for the correct soil initialisation and the observed one (Fig. 9.10).

In the simulation where the two layer scheme was used for initialisation this higher humidity in the boundary leads to an unrealistic enhancement of cloud formation compared to satellite images (Fig. 9.11).

As a consequence of this comparison, it is therefore recommended to choose one of the following configurations for simulations before 2005 with GME driving data:

(1) Usage of the two layer scheme for GME and COSMO model simulations. This means an interpolation from the two layer scheme to the two layer scheme for initialisation or

(2) Usage of newly processed GME analysis data, which is produced by using the multi-layer soil scheme or

(3) Simulation of periods, which are long enough to reduce the humidity by evapotranspiration.

Using newly processed GME analysis data from the years before 2005 one should keep in mind that they can differ considerably from the older ones, which have been obtained by the operational system!

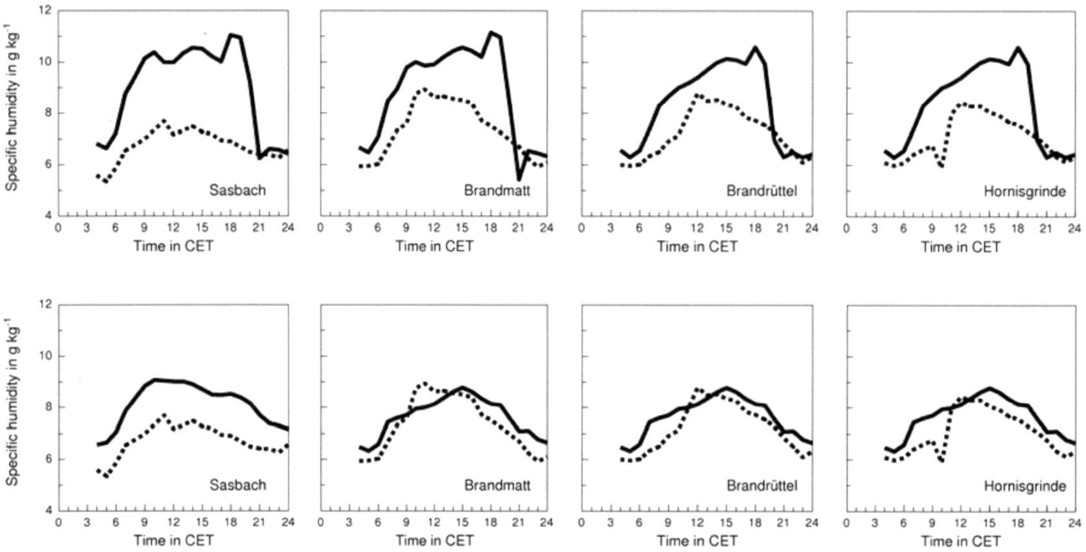

Figure 9.9: Simulations for the first of June 2002 with soil moisture initialisation from the old two layer soil model of GME (top) and from the multi layer scheme of GME (bottom). The observations are indicated by dotted lines and the simulations by solid lines. The location of the observation sites is depicted in Fig. 8.11.

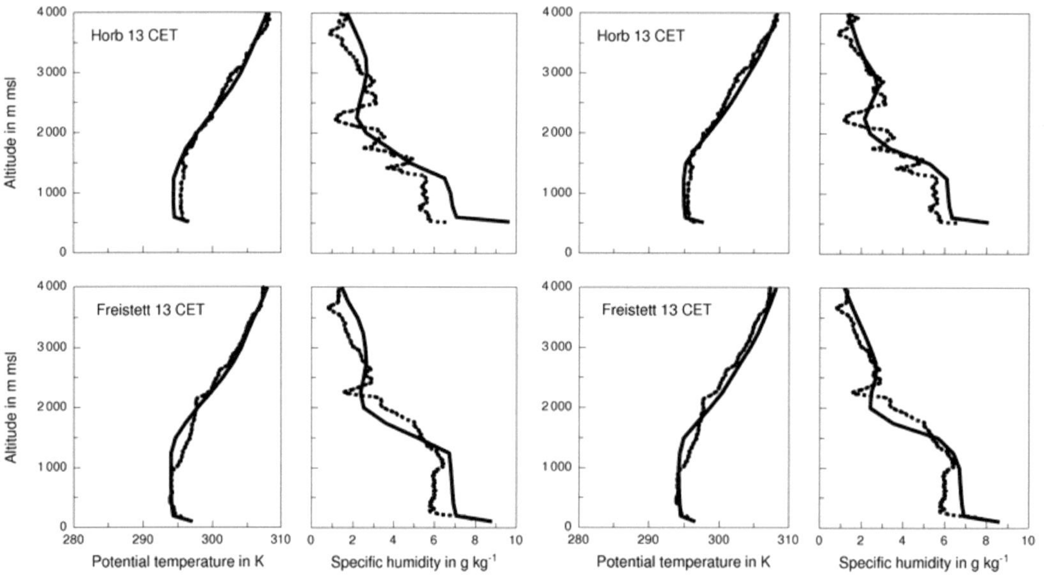

Figure 9.10: Profiles of potential temperature for the first of June 2002 with soil moisture initialisation from the old two layer soil model of GME (left) and from the multi layer scheme of GME (right). The observations are indicated by dotted lines and the simulations by solid lines.

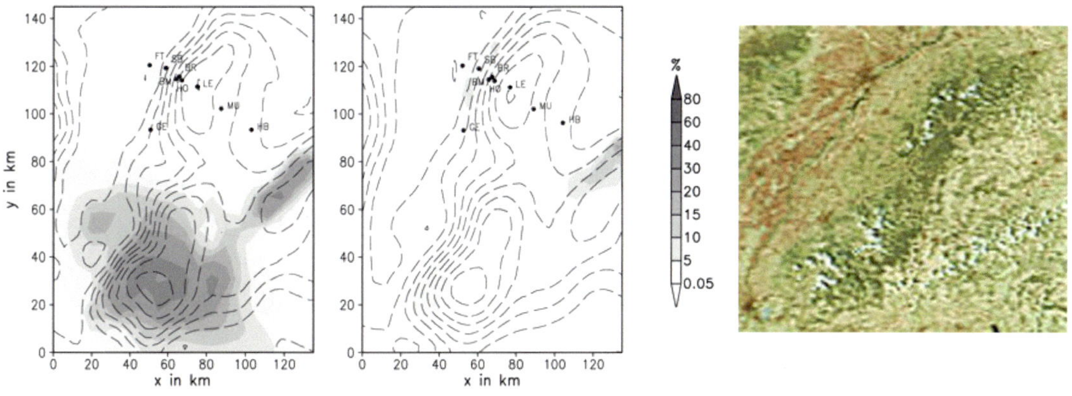

Figure 9.11: Cloud cover for the first of June 2002 in the Black Forest region obtained by the simulation with soil moisture initialisation from the old two layer soil model of GME (left), by the simulation with soil moisture initialisation from the multi layer scheme of GME (middle) and observed by satellite (right). The abbreviations for the observation stations mean FT: Freistett, HB: Horb, SB: Sasbach, BM: Brandmatt, BR: Brandrüttel, HO: Hornisgrinde.

9.4 ABBREVIATIONS

COSMO

LM	**L**okal-**M**odel
CLM	**C**limate version of **LM**
COSMO model	new name for LM
COSMO-CLM	new name for CLM
COSMO-EU	operational COSMO model setup with 7 km grid size

Institutes

DMI	**D**anish **M**eteorological **I**nstitute
DWD	**D**eutscher **W**etter**d**ienst
ECMWF	**E**uropean **C**entre for **M**edium-Range **W**eather **F**orecast
GKSS	**G**esellschaft für **K**ernenergieverwertung in **S**chiffbau und **S**chiffahrt
IMK	**I**nstitut für **M**eteorologie und **K**limaforschung, Universität Karlsruhe
MPI	**M**ax **P**lanck **I**nstitute for Meteorology in Hamburg
NCEP	**N**ational **C**enters for **E**nvironmental **P**rediction
PIK	**P**otsdam **I**nstitut für **K**limafolgenforschung

Global models

GCM	**G**eneral **C**irculation **M**odels
ERA-40	**E**CMWF **R**e-**A**nalysis data
ECHAM5	fifth-generation atmospheric general circulation model developed at MPI in Hamburg
HadCM3	**Had**ley **C**enter **C**oupled **M**odel version **3**

Regional models

RCM	**R**egional **C**limate **M**odel
CRCM	**C**anadian **R**egional **C**limate **M**odel
HIRHAM	Regional climate model developed by MPI and DMI
KAMM	**Ka**rlsruhe **M**esoskaliges **M**odell
MM5	**M**esoscale **M**odel 5
RegCM	**Reg**ional **C**limate **M**odel
REMO	**Re**gional **Mo**del

Land surface schemes

LSS	**L**and **S**urface **S**cheme
SVAT	**S**oil **V**egetation **A**tmosphere **T**ransfere model
BASE	**B**est **a**pproximation of **s**urface **e**xchanges
BATS	**B**ioshere **A**tmosphere **T**ransfer **S**cheme
ISBA	**I**nteraction **S**oil **B**iosphere **A**tmosphere
MOSES	**M**et **O**ffice **S**urface **E**xchange **S**cheme
PILPS	**P**roject of **i**ntercomparison of **l**and-surface **p**arameterization **s**chemes
SiB2	**Si**mple **B**iosphere 2
SNOWMIP2	**Snow** **M**odel **I**ntercomparison **P**roject
TERRA_LM	Soil vegetation model of COSMO
VEG3D	Soil-vegetation model of KAMM

Others

CRU	**C**limate **R**esearch **U**nit
EU	**E**uropean **U**nion
IPCC	**I**ntergovernmental **P**anel on **C**limate **C**hange
KLIWA	**Kli**maveränderung und Konsequenzen für die **Wa**sserwirtschaft
RMSD	**R**oot **M**ean **S**quare **D**ifference
RMSE	**R**oot **M**ean **S**quare **E**rror
SLEVE	**S**mooth **Le**vel **Ve**rtical
SOM	**S**elf-**O**rganizing **M**aps
VERTIKATOR	**Vertika**ler Austausch und **OR**ographie

10 Bibliography

Alley, R., T. Berntsen, N. Bindoff, Z. Chen, A. Chidthaisong, P. Friedlingstein, J. Gregory, G. Hegerl, M. Heimann, B. Hewitson, B. Hoskins, F. Joos, J. Jouzel, V. Kattsov, U. Lohmann, M. Manning, T. Matsuno, M. Molina, N. Nicholls, J. Overpeck, D. Qin, G. Raga, V. Ramaswamy, J. Ren, M. Rusticucci, S. Solomon, R. Somerville, T. Stocker, P. Stott, R. Stouffer, P. Whetton, R. Wood and D. Wratt, 2007. *Climate change 2007: The physical science basis. Summary for Policymakers*. Cambridge University Press, Cambridge, UK.

Anthes, R. A., Y. H. Kuo, R. Y. Hsie, S. Low-Nam and T. W. Bettge, 1989. Estimation skill and uncertainty in regional numerical models. *Quart. J. Roy. Meteorol. Soc.*, **115**, 763–806.

Avissar, R. and R. A. Pielke, 1989. A parameterization of heterogeneous land surfaces for atmospheric numerical models and its impact on regional meteorology. *Mon. Weather Rev.*, **117**, 2113–2136.

Barthlott, C., U. Corsmeier, C. Meißner, F. Braun and C. Kottmeier, 2006: The influence of mesoscale circulation systems on triggering convective cells over complex terrain. *Atmos. Res.*, **81**/2, 150-175.

Beljaars, A. C.M., P. Viterbo, M. J.Miller and A. K. Betts, 1996. The anomalous rainfall over the United States during July 1993: Sensitivity to land surface parameterization and soil moisture anomalies. *Mon. Weather Rev.*, **124**, 362–383.

Block, A., 2007. Unsicherheiten in Oberflächen- und Bodenparametern und ihre Auswirkungen auf die Ergebnisse regionaler Klimasimulationen. PhD thesis, BTU Cottbus.

Bloomsburg, G. L. and S. J. Wang, 1969. Effect of moisture content on permeability of frozen soils. *Paper presented at 16th annual meeting of Pacific Northwest Region, American Geophysical Union*, 11 pp.

Böhm, U., M. Kücken,W. Ahrens, A. Block, D. Hauffe, K. Keuler, B. Rockel and A.Will, 2006. CLM The climate version of LM: Brief Descripition and long-term applications. *COSMO Newsletter*, **6**, 225–235.

Boone, A., V. Masson, T. Meyers and J. Noilhan, 2000. The influence of the inclusion of soil freezing on simulations by a soil-vegetation-atmosphere transfer scheme. *J. Appl. Meteorol.*, **39**, 1544–1569.

Braun, F. J., 2002. Modellierung der Bodenhydrologie. PhD thesis, Fakultät für Physik, Universität Karlsruhe.

Braun, F. J. and G. Schädler, 2005. Comparison of soil hydraulic parameterizations for mesoscale meteorological models. *J. Appl. Meteorol.*, **44**, 1116–1132.

Brooks, R. H. and A. T. Corey, 1964. *Hydraulic properties of porous media.* Hydrol. Pap. 3,Colorado State University, Fort Collins, Colorado.

Campbell, G. S., 1974. A simple method for determining unsaturated conductivity from moisture retention data. *Soil Sci.*, **117**, 311–314.

Caya, D. and R. Laprise, 1999. A semi-implicit semi-Lagrangian regional climate model: the Canadian RCM. *Mon. Weather Rev.*, **127**, 341–362.

Chen, F. and J. Dudhia, 2001. Coupling an advanced land surface-hydrology model with the Penn State-NCAR MM5 modelling system. Part I: Model implementation and sensitivity. *Mon. Weather Rev.*, **129**, 569–585.

Cherkauer, K. A. and D. P. Lettenmaier, 1999. Hydrologic effects of frozen soils in the upper Mississippi River basin. *J. Geophys. Res.*, **104**, 19599–19610.

Christensen, J. H. and O. B. Christensen, 2007. A summary of the PRUDENCE model projections of changes in European climate by the end of this century. *Clim. Change*, **81**, 7–30.

Christensen, J. H., O. B. Christensen, P. Lopez, E. van Meijgaard and M. Botzet, 1996. The HIRHAM4 regional atmospheric climate model. DMI scientific report, Lyngbyvej 100, Copenhagen.

Christensen, O. B., J. H. Christensen, B.Machenhauer and M. Botzet, 1998. Very high-resolution regional climate simulations over Scandinavia - present climate. *J. Clim.*, **11**, 3204–3229.

Cox, P. M., R. A. Betts, C. B. Bunton, R. L. H. Essery, P. R. Rowntree and J. Smith, 1999. The impact of new land surface physics on the GCM simulation of climate and climate sensitivity. *Climate Dyn.*, **15**, 183–203.

Dai, Y. and Coauthors, 2001. *Common Land Model: Technical documentation and users guide*. [Available online at http://climate.eas.gatech.edu/dai/clmdoc.pdf.].

Dalton, F. N. and M. T. van Genuchten, 1986. The time-domain reflectometry method or measureing soil water content and salinity. *Geoderma*, **38**, 237-250.

Davies, H., 1976. A lateral boundary formulation for multi-level prediction models. *Quart. J. Roy. Meteorol. Soc.*, **102**, 405–418.

Deardorff, J. W., 1978. Efficient prediction of ground surface temperature and moisture with inclusion of a layer of vegetation. *J. Geophys. Res.*, **83**, 1889–1904.

Denis, B., R. Laprise and D. Caya, 2003. Sensitivity of a regional climate model to the resolution of the lateral boundary conditions. *Clim. Dyn.*, **20**, 107–126.

Denis, B., R. Laprise, D. Caya and J. Coté, 2002. Downscaling ability of one-way nested regional climate models: the Big-Brother Experiment. *Clim. Dyn.*, **18**, 627–646.

Denmead, O. T. and E. F. Bradley, 1985. *Flux-gradient relationships in a forest canopy*. In: The Forest-Atmosphere Interaction, Eds.: B.A. Hutchinson and B.B. Hicks, Reidel, Dordrecht, 421–442.

Dickinson, R., 1984. Modelling evapotranspiration for three-dimensional global climate models: Climate Processes and Climate Sensitivity. *Geophys. Monograph*, **29**, 58–72.

Dickinson, R., A. Henderson-Sellers and P. Kennedy, 1993. Biosphere Atmosphere Transfer Scheme (BATS) Version 1e as coupled to the NCAR community climate model. NCAR tech.note ncar/tn-3871str, [Available from National Center for Atmospheric Research, P. O. Box 3000 Boulder, CO 80307].

Dickinson, R. E., R.M. Errico, F. Giorgi and G. T. Bates, 1988. Modelling of historic, prehistoric and future climates of the Great Basin. NCAR final rep. to usgs/nnwsi, [Available from NCAR, P.O. Box 3000, Boulder, CO 80307-3000].

Dickinson, R. E., R. M. Errivo, F. Giorgi and G. T. Bates, 1989. A regional climate model for western United States. *Clim. Change*, **15**, 383–422.

Doms, G., J. Förstner, E. Heise, H.-J. Herzog, M. Raschendorfer, R. Schrodin, T. Reinhardt and G. Vogel, 2005. *A description of the non-hydrostatic regional model LM. Part II: physical parameterisation*. Deutscher Wetterdienst, Offenbach, [available at: www.cosmo-model.org].

Doms, G. and U. Schättler, 2002. A description of the nonhydrostatic regional model LM Part I: Dynamics and Numerics. *COSMO Newsletter*, **2**, 225–235.

Douville, H., J.-F. Royer and J.-F. Mahfouf, 1995. A new snow parameterization for the Meteo-France climate model. Part I: Validation in stand-alone experiments. *Climate Dyn.*, **12**, 21–35.

Feldmann, H., B. Früh, G. Schädler, H.-J. Panitz, K. Keuler, M. Lautenschlager and D. Jacob, 2007. Evaluation of RCM precipitation climatologies for south-western Germany. *Met. Zeitschrift, submitted.*

Frei, C., J. H. Christensen, M. Déqué, D. Jacob, R. G. Jones and P. L. Vidale, 2003. Daily precipitation statistics in regional climate models: Evaluation and intercomparison for the European Alps. *J. Geophys. Res.*, **108/D3**, 4124.

Frey-Buness, F., D. Heimann and R. Sausen, 1995. A statistical-dynamical downscaling procedure for global climate simulations. *Theor. Appl. Climatol.*, **50**, 117–131.

Früh, B., J. Bendix, T. Nauss, M. Paulat, A. Pfeiffer, J. W. Schipper, B. Thies and H. Wernli, 2007. Verification of precipitation from regional climate simulations and remote-sensing observations with respect to ground-based observations in the upper Danube catchment. *Met. Zeitschrift*, **16**, 275–293.

Fuentes, U. and D. Heimann, 2000. An improved statistical-dynamical downscaling scheme and its application to the alpine precipitation climatology. *Theor. Appl. Climatol.*, **65**, 119–135.

Förstner, J. and G. Doms, 2004. Runge-Kutta time integration and high-order spatial discretization of advection - A new dynamical core for the LMK. *COSMO Newsletter*, **4**, 198–203.

Georgi, F. and L. O. Mearns, 1991. Approaches to the Simulation of Regional Climate Change:A Review. *Rev. Geophys.*, **29**, 191–216.

Gerstengarbe, F.-W., P. C. Werner and U. Rüge, 1999. Katalog der Grosswetterlagen Europas nach Paul Hess und Helmuth Brezowsky. *Berichte des Deutschen Wetterdienstes*, **113**, 1881–1998.

Giorgi, F. and G. T. Bates, 1989. The climatological skill of a regional model over complex terrain. *Mon. Weather Rev.*, **117**, 2325–2347.

Giorgi, F., R. Marinucci and G. T. Bates, 1993. Development of a second-generation regional climate model (RegCM2): Boundary-layer and radiative transfer processes. *Mon. Weather Rev.*, **121**, 2794–2813.

Giorgi, F. and L. O. Mearns, 1999. Introduction to special section: Regional climate modelling revisited. *J. Geophys. Res.*, **104**, 6335–6352.

Giorgi, F., C. Shields-Brodeur and G. Bates, 1994. Regional climate change scenarios over the United States produced with a nested regional climate model. *J. Clim.*, **9**, 1150–1162.

Grabe, F., 2001. Simulation der Wechselwirkung zwischen Atmosphäre, Vegetation und Erdoberfläche bei Verwendung unterschiedlicher Parametrisierungsansätze. PhD thesis, Fakultät für Physik, Universität Karlsruhe.

Grell, G., J. Dudhia and D. R. Stauffer, 1994. A description of the fifth-generation Penn State/NCAR mesoscale model (MM5). Ncar tech. note ncar/tn-398+str, [Available from National Center for Atmospheric Research, P. O. Box 3000 Boulder, CO 80307].

Günther, T., 2004. *Langzeitverhalten von hydrologischen Größen.* Symposium Klimaveränderung und Konsequenz für die Wasserwirtschaft, Würzburg, 70–89.

HAD, 2003. *Hydrologischer Atlas von Deutschland.* Bundesministerium für Umwelt, Naturschutz und Reaktorsicherheit, Bonn.

Hagemann, S., B. Machenhauer, R. Jones, O. Christensen, M. Déqué, D. Jacob and P. Vidale, 2004. Evaluation of water and energy budgets in regional climate models applied over Europe. *Clim. Dyn.*, **23**, 547–567.

Haller, M., 2005. Simulation einjähriger Klimazeitreihen in Süddeutschland mit dem regionalen Klimamodell CLM. Diploma thesis, Universität Karlsruhe.

Heise, E., M. Lange, B. Ritter and R. Schrodin, 2003. Improvement and validation of the multilayer soil model. *COSMO Newsletter*, **3**, 198–203.

Heret, C., A. Tittebrand and F. Berger, 2006. Latent heat fluxes simulated with a non-hydrostatic weather forecast model using actual surface properties from measurements and remote sensing. *Boundary Layer Meteorol.*, **121**, 175–194.

Hess, R., 2001. Assimilation of screen-level observations by variational soil moisture analysis. *Meteorol. Atmos. Phys.*, **77**, 145–154.

Hewitson, B. C. and R. G. Crane, 2002. Self-organizing maps: applications to synoptic climatology. *Clim. Res*, **22**, 13–26.

Hewitt, C., 2005. Ensemble-based predictions of climate changes and their impacts. *Eos*, **85**, 566.

Hillel, D. J., 1980. *Applications of soil physics*. Academic Press, New York.

Klemp, J. B. and R. Wilhelmson, 1978. The simulation of three-dimensional convective storm dynamics. *J. Atmos. Sci.*, **35**, 1070-1096.

IPCC, 1990. *Climate change - The IPCC scientific assessment.* Camebridge University Press, Camebridge, UK.

IPCC, 2001. *Climate Change: The scientific basis, Intergovernmental Panel on Climate Change*. Camebridge University Press, Camebridge, UK., Camebridge.

Jacob, D., 2001. A note to the simulation of the annual and inter-annual variability of the water budget over the Baltic Sea drainage basin. *Meteorol. Atmos. Phys.*, **77**, 61–73.

Jacob, D., L. Bärring, O. Christensen, J. Christensen, M. de Castro, M. Déqué, F. Giorgi, S. Hagemann, M. Hirschi, R. Jones, E. Kjellström, G. Lenderink, B. Rockel, E. Sánchez, C. Schär, S. Seneviratne, S. Somot, A. van Ulden and B. van den Hurk, 2007. An intercomparison of regional climate models for Europe: model performance in present-day climate. *Clim. Change*, **81**, 31–52.

Jacob, D. and R. Podzun, 1997. Sensitivity studies with the regional climate model REMO. *Meteorol. Atmos. Phys.*, **63**, 119–129.

Johansen, O., 1975. *Thermal conductivity of soils*. Phdthesis, Universität Trondheim.

Jones, R. G., J. M. Murphy and M. Noguer, 1995. Simulation of climate change over Europe using a nested regional-climate model. I: Assessment of control climate, including sensitivity to location of lateral boundaries. *Quart. J. Roy. Meteorol. Soc.*, **121**, 1413–1449.

Jones, R. G., J. M. Murphy, M. Noguer and M. Keen, 1997. Simulation of climate change over Europe using a nested regional climate model. I: Comparison of driving and regional model responses to a doubling of carbon dioxide. *Quart. J. Roy. Meteorol. Soc.*, **123**, 265–292.

Jäger, E. B., I. Anders, D. Lüthi, B. Rockel, C. Schär, H. von Storch and S. I. Seneviratne, 2007. Analysis of ERA40-driven CLM simulations for Europe. *Met. Zeitschrift, submitted.*

Kain, J. S. and J. M. Fritsch, 1990. A one dimensional entraining/detraining plume model and its application to convective parameterization. *J. Atmos. Sci.*, **47**, 2784–2802.

Kanamitsu, M., W. Ebisuzaki, J. Woollen, S.-K. Yang, J. J. Hnilo, M. Fiorino and G. L. Potter, 2002. NCEP-DOE AMIP-II Reanalysis (R-2). *Bull. Am. Meteorol. Soc.*, 1631– 643.

KLIWA, 2005. Langzeitverhalten der Lufttemperatur in Baden-Württemberg und Bayern. Arbeitskreis KLIWA. *KLIWA-Berichte*, **5**, 76 pp.

KLIWA, 2006. Regionale Klimaszenarien für Süddeutschland. Arbeitskreis KLIWA. *KLIWABerichte*, **9**, 100 pp.

Kohonen, T., J. Hynninen, J. Kangas and J. Laaksonen, 1995. *The self-organizing map program package*. Helsinki University of Technology, Finland.

Lenz, C.-J., 1996. Energieumsätze an der Erdoberfläche in gegliedertem Gelände. PhD thesis, Fakultät für Physik, Universität Karlsruhe.

Leung, L. R., L. O. Mearns, F. Giorgi and R. L. Wilby, 2003. Regional climate research. *Bull. Am. Meteorol. Soc.*, **62**, 189–216.

Leung, L. R., Y. Qian, X. Bian, W. M. Washington, J. Han and J. O. Roads, 2004. Mid-century ensemble regional climate change scenarios for the western United States. *Clim. Change*, **62**, 75–113.

Li, X. and T. Koike, 2001. *A new frozen soil parameterization in land surface scheme*. In: Present and Future of Modeling Global Environmental Change: Toward Integrated Modeling. Terrapub, Tokyo, 405–414.

Louis, J.-F., 1979. A parametric model of vertical eddy fluxes in the atmosphere. *Boundary Layer Meteorol.*, **17**, 187–202.

Luo, L., A. Robock, K. Y. Vinnikov, C. A. Schlosser, A. G. Slater, A. Boone, H. Braden, P. Cox, P. de Rosnay, R. Dickinson, Y. Dai, Q. Duan, P. Etchevers, A. Henderson-Sellers, N. Gedney, Y. Gusev, F. Habets, J. Kim, E. Kowalczyk, K. Mitchell, O. Nasonova, J. Noilhan, A. Pitman, J. Schaake, A. Shmakin, T. Smirnova, P. Wetzel, Y. Xue, Z. Yang and Q. Zeng, 2003. Effects of frozen soil on soil temperature, spring infiltration, and runoff: Results from the PILPS 2(d) Experiment at Valdai, Russia. *J. Hydrometeor.*, **4**, 334–351.

Majewski, D., D. Liermann, P. Prohl, B. Ritter, M. Buchhold, T. Hanisch, G. Paul, W. Wergen and J. Baumgardner, 2002. The operational global icosahedral hexagonal gridpoint model GME: Description and high-resolution tests. *Mon. Weather Rev.*, **130**, 319–338.

Manabe, S., 1969. Climate and the ocean circulation. I. The atmospheric circulation and the hydrology of the earth surface. *Mon. Weather Rev.*, **97**, 739–774.

Mayer, H., G. Fernbach, L. Jäger, A. Kessler, A.Matzarakis and D. Redepenning, 2000. Forstmeteorologische Messstelle Hartheim des Meteorologischen Institutes der Universität Freiburg. *Ber. Meteorol. Inst. Univ. Freiburg*, **5**, 55–84.

McGregor, J. L., 1997. Regional climate modelling. *Meteorol. Atmos. Phys.*, **63**, 105–117.

Meißner, C., 2004. Modellierung der Konvektionsentwicklung unter Einfluss unterschiedlicher Oberflächeneigenschaften. Diploma thesis, Universität Karlsruhe.

Mellor, G. L. and T. Yamada, 1974. A hierarchy of turbulence closure models for planetary boundary layers. *J. Atmos. Sci.*, **31**, 1791–1806.

Mo, K., M. Kanamitsu, H.-M. H. Juang and S.-Y. Hong, 2000. Ensemble regional and global climate prediction for the 1997/1998 winter. *J. Geophys. Res.*, **105**, 29609–29623.

Moberg, A. and P. D. Jones, 2004. Regional climate model simulations of daily maximum and minimum near-surface temperatures across Europe compared with observed station data 1961-1990. *Clim. Dyn.*, **23/7-8**, 695–715.

Neisser, J., W. Adam, F. Beyrich, U. Leiterer and H. Steinhagen, 2002. Atmospheric boundary layer monitoring at the Meteorological Observatory Lindenberg as a part of the Lindenberg Column: Facilities and selected results. *Met. Zeitschrift*, **11**, 241–253.

New, M., M. Hulme and P. D. Jones, 2000. Representing twentieth-century space time climate variability. Part II: Development of 1901-1996 monthly grids of terrestrial surface climate. *J. Clim.*, **13**, 2217–2238.

Nobre, P., A. D.Moura and L. Sun, 2001. Dynamical downscaling of seasonal climate prediction over Nordeste Brazil with ECHAM3 and NCEP's regional spectral models at IRI. *Bull. Am. Meteorol. Soc.*, **82**, 2787–2796.

Noguer, M., R. Jones and J. Murphy, 1998. Sources of systematic errors in the climatology of a regional climate model over Europe. *Clim. Dyn.*, **14**, 691–712.
Noilhan, J. and J.-F. Mahfouf, 1996. The ISBA land surface parameterisation scheme. *Glob. Plan. Change*, **13**, 145–159.

Österle, H., F. Gerstengarbe and P. Werner, 2006. Qualitätsprüfung, Ergänzung und Homogenisierung der täglichen Datenreihen in Deutschland, 1951-2003: Ein neuer Datensatz. *In: Proc. 7. Deutsche Klimatagung.*

Panitz, H.-J. 2007. Personal communication. *IMK Karlsruhe.*

Pitman, A. J. and A. Henderson-Sellers, 1998. Recent progress and results from the project for the intercomparison of land surface parameterization schemes. *J. Hydrol.*, **212/213**, 128–135.

Rapp, J. and C.-D. Schönwiese, 1996. *Atlas der Niederschlags- und Temperaturtrends in Deutschland 1891-1990*, Volume **5**. Frankfurter geowissenschaftliche Arbeiten, Serie B.

Rawls, W. and D. Brakensiek, 1982. Estimating soil water retention from soil properties. *J. Irrig. Drain. Div. Am. Soc. Civ. Eng.*, **108**, 166–171.

Rhodin, A., F. Kucharski, U. Callies, D. Eppel and W. Wergen, 1999. Variational analysis of effective soil moisture from screen-level atmospheric parameters: Application to a short-range weather forecast model. *Quart. J. Roy. Meteorol. Soc.*, **125**, 2427–2448.

Richter, A., G. H. Adler, J. Behrens and R. Hartwich, 2003. *Bodenübersicht.* In: Hydrologischer Atlas Deutschland. Bundesministerium für Umwelt, Naturschutz und Reaktorsicherheit, Bonn.

Rijtema, P. E., 1969. Soil moisture forecasting. Technical report no. 513, Instituut voor Cultuurtechnieken Waterhuishouding, Wageningen.

Ritter, B., 2007: Personal communication. *DWD, Offenbach.*

Ritter, B. and J.-F. Geleyn, 1992. A comprehensive radiation scheme for numerical weather prediction models with potential applications in climate simulations. *Mon. Weather Rev.*, **120**, 303–325.

Rodell, M., P. R. Houser, A. A. Berg and J. S. Famiglietti, 2005. Evaluation of 10 methods for initializing a land surface model. *J. Hydromet.*, **6**, 146–155.

Rodriguez-Iturbe, I., D. Entekhabi and R. L. Bras, 1991. Nonlinear dynamics of soil moisture at climate scales. 1. Stochastic Analysis. *Water Ress. Res.*, **27**, 1899–1906.

Rojas, M. and A. Seth, 2003. Simulation and sensitivity in a nested modelling system for South America. Part II: GCM boundary forcing. *J. Clim.*, **16**, 2454–2471.

Rost, J. and H. Mayer, 2006. Comparative analysis of albedo and surface energy balance of a grassland site and an adjacent Scots pine forest. Clim. Res., **30**, 227–237.

Sasse, R., 2004. Beziehung zwischen Niederschlag und synoptisch-skaligen Prozessen im östlichen Mittelmeergebiet. Diploma thesis, Universität Karlsruhe.

Schädler, G., 1990. Numerische Simulationen zur Wechselwirkung zwischen Landoberflächen und atmosphärischer Grenzschicht. PhD thesis, Fakultät für Physik, Universität Karlsruhe.

Schädler, G., N. Kalthoff and F. Fiedler, 1990. Validation of a model for heat, mass and momentum exchange over vegetated surfaces using LOTREX-10E/HIBE88 data. *Beitr. Phys. Atmosph.*, **63**, 85–100.

Schädler, G. and R. Sasse, 2006. Analysis of the connection between precipitation and synoptic scale processes in the Eastern Mediterranean using self-organizing maps. *Met. Zeitschrift*, **15/3**, 273–278.

Schär, C., D. Lüthi, U. Beyerle and E. Heise, 1999. The Soil Precipitation Feedback: A process study with a regional climate model. *J. Clim.*, **12**, 722–741.

Seifert, A., 2007. A revised cloud microphysical parameterization for COSMO-LME. *WGNE Blue Book, http://www.cmc.ec.gc.ca/rpn/wgne.*

Seifert, A. and K. D. Beheng, 2001. A double moment parameterization for simulating autoconversion and selfcollection. *Atmos. Res.*, **59-60**, 265–281.

Seth, A. and F. Giorgi, 1998. The effects of domain choice on summer precipitation. Simulation and sensitivity in a regional climate model. *J. Clim.*, **11**, 2698–2712.

Simmons, A., P. Jones, V. da Costa Bechtold, A. Beljaars, P. Kallberg, S. Saarinen, S. Uppala, P. Viterbo and N. Wedi, 2004. Comparison of trends and low-frequency variability in CRU, ERA-40, and NCEP/NCAR analyses of surface air temperature. *J. Geophys. Res.*, **109**, D24115.

Simmons, A. J. and J. K. Gibson, 2000. The ERA-40 Project Plan. *ERA-40 Project Report Series*, **1**, 62 pp.

Skamarock, W. C. and J. B. Klemp, 1992. The stability of time-split numerical methods for the hydrostatic and the nonhydrostatic elastic equations. *Mon. Wather Rev.*, **120**, 2109-2127.

Slater, A. G., A. J. Pitman and C. E. Desborough, 1998. Simulation of freeze-thaw cycles in a general circulation model land surface scheme. *J. Geophys. Res.*, **103**, 11303–11312.

Smith, C. B., M. N. Lakhtakia, W. J. Capehart and T. N. Carlson, 1994. Initialization of soilwater content in regional-scale atmospheric prediction models. *Bull. Am. Meteorol. Soc.*, **75**, 585–593.

Smoydzin, L., 2004. Theoretische und numerische Untersuchungen zur Konvektionsparametrisierung in einem Wettervorhersagemodell. Diploma thesis, Universität Bonn.

Straub, H., 2004. *Langzeitverhalten von hydrologischen Größen.* KLIWA Berichte 4/2. Symposium Klimaveränderung und Konsequenz für die Wasserwirtschaft, Würzburg, 57–72.

Stull, R., 1988. *Boundary Layer Meteorology*. Kluwer Academic.

Taconet, O., R. Bernard and D. Vidal-Madjar, 1986. Evapotranspiration over an agricultural region using a surface flux/temperature model based on NOAA-AVHRR data. *J. Appl. Meteorol.*, **25**, 284–307.

Tiedtke, M., 1989. A comprehensive mass flux scheme for cumulus parameterization in largescale models. *Mon. Weather Rev.*, **117**, 1779–1800.

Timbal, B. and A. Henderson-Sellers, 1998. Intercomparisons of land-surface parameterizations coupled to a limited area forecast model. *Glob. Plan. Change*, **19**, 247–260.

Van Genuchten, M., 1980. A closed-form equation for predicting the hydraulic conductivity of unsaturated soils. *Soil. Sci. Soc. Am. J.*, **44**, 892–898.

Verseghy, D., 1991. CLASS-A Canadian land surface scheme for GCMS. I. soil model. *Int. J. Climatology*, **11**, 111–133.

Vidale, P.-L., D. Lüthi, C. Frei, S. Seneviratne and C. Schär, 2003. Predictability and uncertainty in a regional climate model. *J. Geophys. Res.*, **108**, 4586.

Viterbo, P., A. Beljaars, J.-F. Mahfouf and J. Teixeira, 1999. The representation of soil moisture freezing and its impact on the stable boundary layer. *Quart. J. Roy. Meteorol. Soc.*, **125**, 2401–2426.

Von Storch, H., H. Langenberg and F. Feser, 2000. A spectral nudging technique for dynamical downscaling purposes. *Mon. Weather Rev.*, **128**, 3664–3673.

Wang, Y., L. R. Leung, J. L. McGregor, D.-K. Lee, W.-C. Wang, Y. Ding and F. Kimura, 2004. Regional climate modeling: Progress, challenges, and prospects. *J. Meteorol. Soc. Jap.*, **82/6**, 1599–1628.

Wang, Y., O. L. Sen and B. Wang, 2003. A highly resolved regional climate model (IPR-CRegCM) and its simulation of 1998 severe precipitation event over China. Part I: Model Description and verification of simulation. *J. Clim.*, **16**, 1721–1738.

Warner, T. T., R. A. Peterson and R. E. Treadon, 1997. A tutorial on lateral boundary conditions as a basic and potentially serious limitation to regional numerical weather prediction. *Bull. Am. Meteorol. Soc.*, **78**, 2599–2617.

Wilby, R. L., C. W. Dawson and E. M. Barrow, 2002. SDSM - a decision support tool for the assessment of regional climate change impacts. *Environmental Modelling and Software*, **17**, 147–159.

Will, A., 2005. Personal communication. *BTU Cottbus*.

Will, A. and Coauthors, 2007. Physics and dynamics of the CLM. *Met. Zeitschrift, submitted.*

Willmott, C. and K. Matsuura, 2001. Terrestrial air temperature and precipitation: Monthly and annual time series 1950-1999. [available at: http//:climate.geog.udel.edu].

Wohnsiedler, S., 2006. Erstellung und Einsatz eines Bodenkatasters für regionale Klimasimulationen. Diploma thesis, Universität Karlsruhe.

Yang, Z.-L., R. E. Dickinson, A. Henderson-Sellers and A. J. Pitman, 1995. Preliminary study of spin-up processes in land surface models with the first stage data of Project for Intercomparison of Land Surface Parameterization Schemes Phase 1(a). *J. Geophys. Res.*, **100/D8**, 16553–16578.

Yu, Z. and L. Shihua, 2002. Development and validation of a simple frozen soil parameterization scheme used for climate model. *Earth Environm. Sci.*, **19/3**, 513–527.

Zhao, T. and C. Fu, 2006. Comparison of products from ERA-40, NCEP-2, and CRU with station data for summer precipitation over China. *Adv. Atmos. Sci.*, **23**, 593–604.

Danksagung

Die vorliegende Arbeit wurde am Institut für Meteorologie und Klimaforschung des Forschungszentrums Karlsruhe und der Universität Karlsruhe durchgeführt. Die notwendigen Finanz- und Sachmittel wurden vom Forschungszentrum Karlsruhe zur Verfügung gestellt.

Herrn Prof. Dr. Ch. Kottmeier danke ich sehr für die interessante Themenstellung und die fortwährende Unterstützung meiner Arbeit.

Frau Prof. Dr. S. Jones möchte ich für die Übernahme des Korreferates sehr herzlich danken.

Mein besonderer Dank gilt Herrn Dr. Gerd Schädler für die Betreuung dieser Arbeit, seine stete Unterstützung und die interessanten Diskussionen.

Danken möchte ich an dieser Stelle auch all denjenigen, die mir Daten oder Programme zur Verfügung gestellt haben und mich in allen Fragen diesbezüglich unterstützt haben. Dies waren im einzelnen: Burkhardt Rockel (GKSS Geesthacht), Felix Ament (Meteo Schweiz), Daniel Lüthi (ETH Zürich), Herr Österle (PIK Potsdam), Gerd Vogel (Observatorium Lindenberg), Dirk Schindler (Universität Freiburg), Axel Seifert (DWD), Jochen Förstner (DWD), Ulrich Schättler (DWD), Andreas Will (BTU Cottbus), Hans-Jürgen Panitz, Barbara Früh und alle, die ich vielleicht vergessen haben sollte.

Dem Höchstleistungsrechenzentrum Stuttgart danke ich für die Bereitstellung der Rechenmöglichkeiten, ohne die diese Arbeit nicht möglich gewesen wäre.

Die NCEP-DOE Reanalyse Daten wurden bereitgestellt von NOAA/OAR/ESRL PSD, Boulder, Colorado, USA (http://www.cdc.noaa.gov/) und die ERA-40 Daten wurden bereitgestellt vom Europäischen Zentrum für mittel- und langfristige Wettervorhersage (from http://www.ecmwf.int/).

Danken möchte ich auch all denjenigen, die für meine Fragen und Probleme bezüglich dieser Arbeit und darüber hinaus stets ein offenes Ohr und gute Antworten hatten, insbesondere den Mitarbeitern der 5. Ebene.

Gedankt sei auch all denjenigen, die sich die Zeit nahmen, diese Arbeit zu lesen, und wertvolle Korrekturvorschläge beisteuerten.

Meiner Zimmerkollegin Tanja Stanelle möchte ich an dieser Stelle ebenfalls danken für die schöne Zeit und ich glaube, dass uns die vielen Tage und Nächte in netter Atmosphäre lange in Erinnerung bleiben werden.

Meinem Freund danke ich für die vielen Stunden, die er mit Korrekturlesen verbracht hat und für die Aufmunterungen aus der Ferne.

Zuletzt möchte ich meinen Eltern danken, die mein Studium ermöglicht haben und die auch während meiner Promotionszeit immer für mich da waren.

Wissenschaftliche Berichte des
Instituts für Meteorologie und Klimforschung der Universität Karlsruhe

Bisher erschienen:

Nr. 1: Fiedler,F., Prenosil, T.:
Das MESOKLIP-Experiment.
(Mesoskaliges Klimaprogramm im Oberrheintal).
August 1980

Nr. 2: Tangermann-Dlugi, G.:
Numerische Simulationen atmosphärischer Grenzschichtströmungen über langgestreckten mesoskaligen Hügelketten bei neutraler thermischer Schichtung.
August 1982

Nr. 3: Witte, N.:
Ein numerisches Modell des Wärmehaushalts fließender Gewässer unter Berücksichtigung thermischer Eingriffe.
Dezember 1982

Nr. 4: Fiedler, F. und Höschele, K. (Hrsg.):
Prof. Dr. Max Diem zum 70. Geburtstag.
Februar 1983 (vergriffen)

Nr. 5: Adrian, G.:
Ein Initialisierungsverfahren für numerische mesoskalige Strömungsmodelle.
Juli 1985

Nr. 6: Dorwarth, G.:
Numerische Berechnung des Druckwiderstandes typischer Geländeformen.
Januar 1986

Nr. 7: Vogel, B., Adrian, G., Fiedler, F.:
MESOKLIP-Analysen der meteorologischen Beobachtungen von mesoskaligen Phänomenen im Oberrheingraben.
November 1987

Nr. 8: Hugelmann, C.-P.:
Differenzenverfahren zur Behandlung der Advektion.
Februar 1988

Nr. 9: Hafner, T.:
Experimentelle Untersuchungen zum Druckwiderstand der Alpen.
April 1988

Nr. 10: Corsmeier, U.:
Analyse turbulenter Bewegungsvorgänge in der maritimen atmosphärischen Grenzschicht.
Mai 1988

Nr. 11: Walk, O. and Wieringa, J.(eds):
Tsumeb Studies of the Tropical Boundary-Layer Climate.
Juli 1988

Nr. 12: Degrazia, G. A.:
Anwendung von Ähnlichkeitsverfahren auf die turbulente Diffusion in der konvektiven und stabilen Grenzschicht.
Januar 1989

Nr. 13: Schädler, G.:
Numerische Simulationen zur Wechselwirkung zwischen Landoberflächen und atmophärischer Grenzschicht.
November 1990

Nr. 14: Heldt, K.:
Untersuchungen zur Überströmung eines mikroskaligen Hindernisses in der Atmosphäre.
Juli 1991

Nr. 15: Vogel, H.:
Verteilungen reaktiver Luftbeimengungen im Lee einer Stadt -
Numerische Untersuchungen der relevanten Prozesse.
Juli 1991

Nr. 16: Höschele, K.(ed.):
Planning Applications of Urban and Building Climatology -
Proceedings of the IFHP / CIB-Symposium Berlin, October 14 - 15, 1991.
März 1992

Nr. 17: Frank, H.P.:
Grenzschichtstruktur in Fronten.
März 1992

Nr. 18: Müller, A.:
Parallelisierung numerischer Verfahren zur Beschreibung von Ausbreitungs- und chemischen Umwandlungsprozessen in der atmosphärischen Grenzschicht.
Februar 1996

Nr. 19: Lenz, C.-J.:
Energieumsetzungen an der Erdoberfläche in gegliedertem Gelände.
Juni 1996

Nr. 20: Schwartz, A.:
Numerische Simulationen zur Massenbilanz chemisch reaktiver Substanzen im mesoskaligen Bereich.
November 1996

Nr. 21: Beheng, K.D.:
Professor Dr. Franz Fiedler zum 60. Geburtstag.
Januar 1998

Nr. 22: Niemann, V.:
Numerische Simulation turbulenter Scherströmungen mit einem Kaskadenmodell.
April 1998

Nr. 23: Koßmann, M.:
Einfluß orographisch induzierter Transportprozesse auf die Struktur der atmosphärischen Grenzschicht und die Verteilung von Spurengasen.
April 1998

Nr. 24: Baldauf, M.:
Die effektive Rauhigkeit über komplexem Gelände - Ein Störungstheoretischer Ansatz.
Juni 1998

Nr. 25: Noppel, H.:
Untersuchung des vertikalen Wärmetransports durch die Hangwindzirkulation auf regionaler Skala.
Dezember 1999

Nr. 26: Kuntze, K.:
Vertikaler Austausch und chemische Umwandlung von Spurenstoffen über topographisch gegliedertem Gelände.
Oktober 2001

Nr. 27: Wilms-Grabe, W.:
Vierdimensionale Datenassimilation als Methode zur Kopplung zweier verschiedenskaliger meteorologischer Modellsysteme.
Oktober 2001

Nr. 28: Grabe, F.:
Simulation der Wechselwirkung zwischen Atmosphäre, Vegetation und Erdoberfläche bei Verwendung unterschiedlicher Parametrisierungsansätze.
Januar 2002

Nr. 29: Riemer, N.:
Numerische Simulationen zur Wirkung des Aerosols auf die troposphärische Chemie und die Sichtweite.
Mai 2002

Nr. 30: Braun, F. J.:
Mesoskalige Modellierung der Bodenhydrologie.
Dezember 2002

Nr. 31: Kunz, M.:
Simulation von Starkniederschlägen mit langer Andauer über Mittelgebirgen.
März 2003

Nr. 32: Bäumer, D.:
Transport und chemische Umwandlung von Luftschadstoffen im Nahbereich von Autobahnen - numerische Simulationen.
Juni 2003

Nr. 33: Barthlott, C.:
Kohärente Wirbelstrukturen in der atmosphärischen Grenzschicht.
Juni 2003

Nr. 34: Wieser, A.:
Messung turbulenter Spurengasflüsse vom Flugzeug aus.
Januar 2005

Nr. 35: Blahak, U.:
Analyse des Extinktionseffektes bei Niederschlagsmessungen mit einem C-Band Radar anhand von Simulation und Messung.
Februar 2005

Nr. 36: Bertram, I.:
Bestimmung der Wasser- und Eismasse hochreichender konvektiver Wolken anhand von Radardaten, Modellergebnissen und konzeptioneller Betrachtungen.
Mai 2005

Nr. 37: Schmoeckel, J.:
Orographischer Einfluss auf die Strömung abgeleitet aus Sturmschäden im Schwarzwald während des Orkans „Lothar".
Mai 2006

Nr. 38: Schmitt, C.:
Interannual Variability in Antarctic Sea Ice Motion.
Interannuelle Variabilität antarktischer Meereis-Drift.
Mai 2006

Nr. 39: Hasel, M.:
Strukturmerkmale und Modelldarstellung der Konvektion über Mittelgebirgen
Juli 2006

Ab Band 40 erscheinen die Wissenschaftlichen Berichte des Instituts für Meteorologie und Klimaforschung im Universitätsverlag Karlsruhe online unter der Internetadresse: http://www.uvka.de/
Auf Wunsch sind beim Universitätsverlag Karlsruhe auch gedruckte Exemplare erhältlich („print on demand").

Nr. 40: Lux, R.:
Modellsimulationen zur Strömungsverstärkung von orographischen Grundstrukturen bei Sturmsituationen.
Mai 2007

Nr. 41: Straub, W.:
Der Einfluss von Gebirgswellen auf die Initiierung und Entwicklung konvektiver Wolken.
März 2008

Nr. 42: Meißner, C.:
High-resolution sensitivity studies with the regional climate model COSMO-CLM.
Dezember 2008

Made in the USA
Las Vegas, NV
03 November 2026

57425687R00126